AMBER

Window to the Past

AMBER

Window to the Past

David A. Grimaldi

Harry N. Abrams, Inc., Publishers, in association with

the American Museum of Natural History

TO THE LADIES IN MY LIFE:
KAREN, REBECCA, AND LITTLE EMILY

Page 2: True scorpion. Dominican amber, 3 x 4". Private collection

Page 6: Chest. Replica of seventeenth-century German design, made c. 1880 in Munich for Arnold Buffum by Fritz von Müller, director of the Academy of Art in Munich. Height 18". Courtesy, Museum of Fine Arts, Boston. Bequest of William Arnold Buffum, 02.86

The body of the chest is of ebony and silver gilt, with "windows" of transparent amber medallions, cut in profiles and portraits, inserted. The amber has been called Sicilian but is most likely Baltic.

Page 7: Man carrying a burden. China, eighteenth century or earlier. Height 3.2". American Museum of Natural History (Anthropology). Drummond Collection, 70.3.2584

The figure is carved from a single piece of clear yellow amber, the base from another piece of similar color.

Editor: Harriet Whelchel
Designer: Maria Learmonth Miller

Library of Congress Cataloging-in-Publication Data

Grimaldi, David A.
    Amber : window to the past / by David A. Grimaldi.
        p.   cm.
    Includes bibliographical references and index.
    ISBN 0–8109–1966–4 (Abrams: cloth) / 0–8109–2652–0 (Mus. pbk)
    1. Amber.   2. Amber art objects.   I. Title.
    QE391.A5G76   1996
    553.2'9—dc20
                                                    95–651

Printed and bound in Japan

CONTENTS

PREFACE AND ACKNOWLEDGMENTS

*B*eing a scientist had been a goal of mine ever since I was a child, but it was not until I was an undergraduate that I realized one could actually do science for a living (albeit the main reward being personal, not at all monetary). I was interested in all aspects of natural history, and deeply so in fossils and insects. In my first year as a graduate student at Cornell, I met Jake Brodzinsky, a noted dealer of insects fossilized in Dominican amber, who showed me the variety of life preserved in that substance. All other fossils just seemed rendered flat in rocks. The fascination has grown ever since, and I have ardently collected amber in various parts of North and Central America (including the Caribbean). As I learned more about amber, I gradually came to realize how few specialists there are about amber in general. As a museum curator, and somewhat out of necessity, I have been involved in studies on the chemistry, paleontology, and provenance of amber, and it has been a delightfully eclectic pursuit.

The inspiration for this book came from a desire to produce a lavishly illustrated volume on the entire spectrum of amber. The text, of course, should be accurate and informative, but the images should speak for themselves. A high standard for photographs was set by Dieter Schlee at the Museum für Naturkunde in Stuttgart. These beautiful photographs, which were published in the Stuttgarter Beiträge series, are of wondrous fossil and mineralogical pieces of amber, but the booklets are in German and not easily obtained. When the American Museum of Natural History launched an exhibit on the natural history and artistry of amber, opening in March 1996, it was a prime opportunity to produce such a book, which would be a guide for developing as well as enjoying the exhibit.

Various books written on amber fulfill specific needs. The general books by Patty Rice and Helen Fraquet have texts that are well researched and informative. Likewise, there are several academic books on the paleontology of amber, such as the ones by Sven Larsson and George Poinar. Yet, something was still needed to kindle the popular imagination vis-à-vis captivating images. Several scholarly works on amber in art are informative but are very focused on specific collections, such as Marjorie Trusted's catalogue of the collection of European ambers in the Victoria and Albert Museum, D. E. Strong's catalogue of the ancient ambers in the British Museum, and Alfred Rohde's great 1937 classic on the eighteenth-century European decorative arts. Perhaps the closest equivalent to the present volume is Gisela Reineking von Bock's 1981 book, but it has more black-and-white than color photographs, is mostly about European decorative objects, and is available only in German.

As mediums both for objets d'art and the preservation of extinct organisms with unparalleled fidelity, amber and resins fall into their own category of substances. Science can sometimes reduce the mystique of a subject. In the case of amber, current scientific inquiry has actually added more romance to an already mystical substance.

An American fascination with amber has been fueled by various scientific discoveries that have been widely popularized, many of them having been made at the American Museum of Natural History. It is my hope that the present book and the exhibit will help to make that fascination grow.

A book like this cannot be developed without the help and cooperation of many talented people. It is a pleasure to thank the following individuals for their help, especially (at the American Museum of Natural History) Denis Finnin and Jacklyn Beckett in the Photography Studio and President and Chairman Emeritus of the AMNH, Robert Goelet, for his personal generosity in sponsoring amber research; and numerous others who helped in a great variety of ways: Dr. Herbert Axelrod, Ed Bridges, Joe Peters, Sarah Covington, Joel Sweimler, Don Clyde, Barbara Conklin, Sam Taylor, Gerard Case, Julian Stark, Linda Krause, Hank Silverstein, Lisa Stillman, Donna Englund, and Bea Brewster. Without the support of and talent at the Museum this work would have been much more difficult. Much of my scientific research on amber has been generously sponsored by a grant from the National Science Foundation.

I am also indebted to people at other institutions, and various private individuals, especially Ettore and Remo Morone, for their gracious support and help in studying their wondrous collection; Dieter Schlee (Museum für Naturkunde, Stuttgart); Susan Hendrickson; Alexander Shedrinsky (New York University Institute of Fine Arts); and the conservators and photographers at the Museum of Fine Arts, Boston, who worked very hard to prepare the Buffum Collection for this book and the exhibition. The support of Dr. Anne Poulet and Janis Staggs at the Museum of Fine Arts is deeply appreciated.

There are numerous others who arranged for loans or contributed information: Faya Causi (Washington D.C.); Andrew Ross, Richard Fortey, and Andrew Clark (Natural History Museum, London); John Cooper (Booth Museum, Brighton); Ivan Sautov (Ekaterininsky Palace Museum, St. Petersburg); Cristina Piacenti (Museo degli Argenti, Florence); Marjorie Trusted (Victoria and Albert Museum, London); Vladimir Zherikhin and Yuri Popov (Paleontological Institute, Moscow); William Crepet and Rudolf Meier (Cornell University); Judith and Michael Steinhardt; James Watt, Joan Mertens, and Claire Vincent (Metropolitan Museum of Art); Susana Pancaldo, Shelby White, and Leon Levy; and Laura Siegel (Robert Haber Gallery, New York).

To all I owe deep thanks.

Overleaf: Portions of three contour or flight feathers. Length of amber 1.5". Private collection

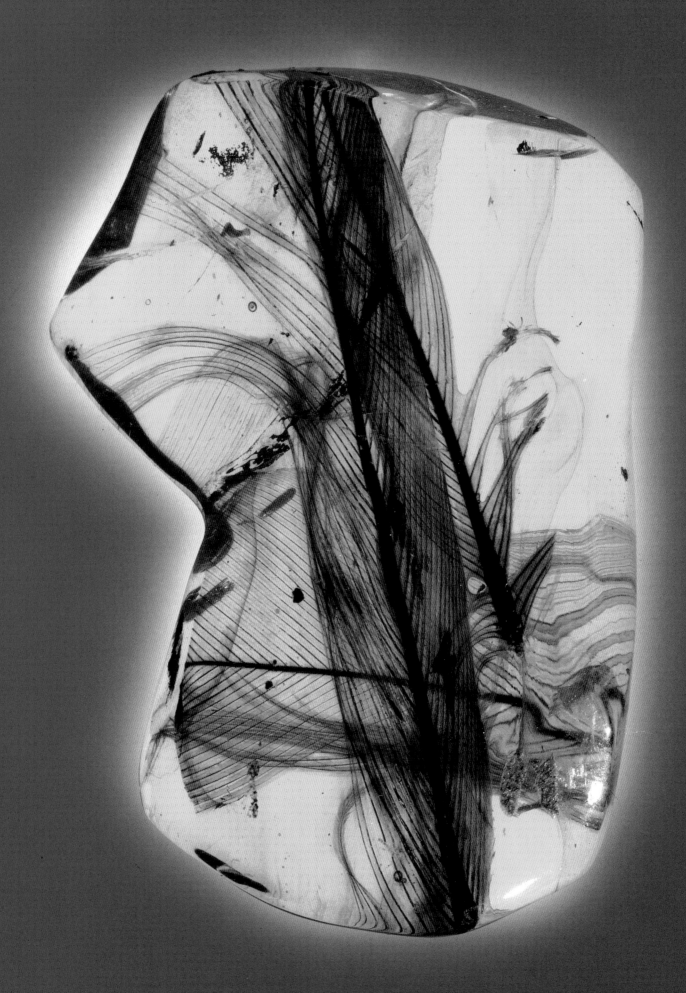

AMBER

IN

NATURE

ORIGINS AND PROPERTIES

*T*he word *amber* can have many associations. It is not a mineral but is used as and called a semiprecious stone. The oldest and most continuous use of it, in fact, is for adornment. Although it is ancient tree resin, amber is not exactly fossilized. We often think of fossils as being the remains of extinct organisms, like dinosaur bones, and impressions of ferns, leaves, and insect wings in rocks. Unlike these kinds of fossils, which are usually mineral replacements of the original structure, amber is entirely organic; its composition from the original resin has changed little over millions of years. Even the inclusions of tiny organisms in amber are strikingly intact. The most common response from people who have seen their first amber fossil is one almost of disbelief that something so old could be so beautifully preserved. Exquisite preservation is a natural property of certain kinds of resins, although the process is not understood very well.

Hundreds of deposits of amber occur around the world, most of them in trace quantities. One would find amber in any place where the hardened resin of various extinct plants would be preserved, but special conditions are required to preserve this substance over millions of years, and only occasionally has amber survived in quantities large enough to be mined. There exist only about twenty such rich deposits of amber in the world, and the deposits vary greatly in age. It is a common misconception that amber is exclusively the fossilized resin of pines; in fact, amber was formed by various conifer trees (only a few of them apparently related to pines), as well as by some tropical broad-leaved trees. Origins of specific amber deposits are presented in detail later in this book.

Most deposits of amber are in marine sediments. Buoyant in water, resin would have floated down rivers with logs and fallen trees and eventually become stranded and concentrated on the shores. Sediments gradually covered the hardened resin, logs, and branches. Over thousands to millions of years, the wood became lignite and the resin turned to amber.

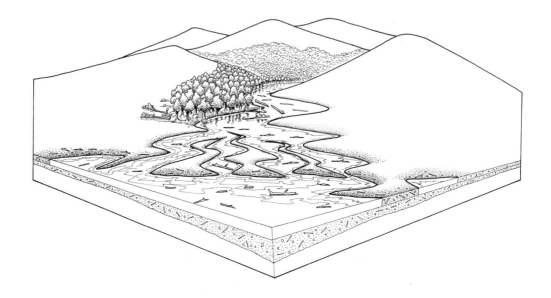

	Evolution of hominids	
	Quaternary	**2.5**
	Pliocene	**7**
	Bitterfeld amber	
	Miocene	**26**
	Dominican, Mexican ambers	
	Oligocene	**38**
	Baltic, Arkansas ambers	
	Eocene	**54**
	Sakhalin amber	
	Modern orders of mammals	
TERTIARY	Paleocene	**65**
	Extinctions of dinosaurs	
	Radiations of flowering plants	
	Modern families of insects	
	Amber from Canada, France, Lebanon, New Jersey, Siberia	
CRETACEOUS		**141**
	Birds evolve	
	Dinosaurs diversify	
	Conifers dominate	
JURASSIC		**195**
	Cycads and gingkos evolve	
	Ferns and conifers diversify	
	rise of dinosaurs	
TRIASSIC		**225 million years ago**

Geological time scale, showing the periods relevant to the formation of amber

Amber is almost always preserved in a sediment that formed the bottom of an ancient lagoon or river delta at the edge of an ocean or sea. The specific gravity of solid amber is only slightly higher than that of water; although it does not float, it is buoyant and easily carried by water (amber with air bubbles is even more buoyant). Thus, amber would be carried downriver with logs from fallen amber trees and cast up as beach drift on the shores or in the shallows of the delta into which the river empties. Over time, sediments would gradually bury the wood and resin. The resin would become amber, and the wood a blackened, charcoal-like lignite.

Amber is often preserved this way because, unless protected, the surface of amber reacts with oxygen in the atmosphere and, over many years, develops an opaque crust. Given enough time, the entire piece will crumble away. Dense, wet sediments of clay and sand are virtually devoid of oxygen and preserve amber extremely well. Today, most amber is found by searching for lignite in the sandstones, loose shales, and mudstone remains of deltaic sediments. A rich amber deposit is thus a combination of fortuitous factors involving concentration of the amber, appropriate burial, as well as a forest nearby that produced large quantities of the right kind of resin.

Resins

It is ironic that such a beautiful and mystical substance as amber is also one that is probably just a by-product of trees defending themselves against insects and disease. Some trees, like ponderosa pine, produce massive flows of resin when bark beetles chew galleries into the wood. Living relatives of the various amber trees, especially from the tropics, are copious resin producers. Perhaps this is related to more intensive insect attacks in the tropics, but insect attacks are not the only causes of resin production: heavy flows are also caused by wounds, such as a snapped tree limb or gashed trunk. As the resin wells to the surface, it covers the wound and hardens, thus acting as a seal against further damage by fungi and insects. But before it hardens, small insects, spiders, and even tiny vertebrates may become mired in the resin and, eventually, encapsulated and mummified. Presumably, the same chemical features of the resin that preserve it over millions of years are the ones that preserve the tiny organisms in it with such fidelity.

Resins vary tremendously in composition and have different fragrances and colors, but they all have *terpenes,* which are the compounds that become linked as the resin hardens into amber. Some terpenes are very volatile and dissipate quickly into the air as the resin hardens; others remain as a virtually inert part of the amber. It is the "bouquet" of various terpenes that renders the distinctive aromas of resins and ambers.

The special properties of resins have been recognized and exploited for thousands of years. Ancient Egyptians used sandarac (from *Callitris* and *Tetraclinus* trees) and mastic (from *Pistacea*) as a base for pigments that were painted onto jars and the walls of tombs. The great masters hardened their oil paints and coated their paintings with liquid dammar (a generic Malay word for all resins but generally used to refer to those from Southeast Asia). Varnishes and lacquers were produced from ground copal and amber.

Resins and amber were also surrounded by a rich medicinal mystique. Some native North Americans used resins from cedars, firs, and pines for various ailments. The Maya even medicinally used resin from *Hymenaea* trees, which we now know is very similar to the amber from Mexico and the Dominican Republic. John Cook, M.D., prescribes in his 1770 treatise, *The Natural History of Lac, Amber, and Myrrh:*

> *Many are the excellent virtues of Amber, especially when taken inwardly, in a cold state of the Brain, in Catarrhs, in the Headache, sleep and convulsive disorders, in the suppression of the menses, hysterical and hypochondriacal disorders, and in hemorrhages or bleedings.*

Cook's recommended dosage was "60 or 80 drops for grown persons, two or three times a day, in any liquid." Cadawallader Colden, a distinguished colonial physician in America, extolled the virtues of an unlikely concoction of ground pine resin steeped in water, called "Tar Water." Several seventeenth-century treatises were written on this odd cordial as a treatment against smallpox, ulcers, diarrhea, and the "foulest distempers" (syphilis). (In an age like ours, in which folk medicine is revealing a wealth of medical insights from tropical plants, such remedies should not be immediately dismissed.) Few substances, though, rival the mystical powers of the most famous resins, frankincense and myrrh.

Frankincense is the resin from *Boswellia* trees, especially the species *carterii, papyrifera,* and *thurifera.* The finest frankincense and perhaps the oldest harvests are from southern Arabia. From here the Hadramis would transport the material via camel caravans across the Arabian sands to Palestine and Egypt, and other merchants would bring it to Greece and Rome, where it was especially prized. In the second century A.D., 3,000 tons per year were shipped throughout the Mediterranean, most of it to the Romans. Its value to the people of Palestine is reflected in its mention in the Bible twenty-two times. Its extremely rich, resinous aroma made it the finest incense available, and it was burned (sometimes continuously) in temples and even used as a base for perfumes. The value of frankincense vied with that of gold; it was offered to the infant Christ by the Magi along with gold and myrrh (Matthew 2:11).

Myrrh is from shrubby *Commiphora* trees, which are found in the same regions as are *Boswellia.* Myrrh, too, was used as an incense, particularly during cremation, and as a base for perfumes, even as the anointing oil of the Hebrews in the Old Testament: "Your God has set you above your companions, by anointing you with the oil of joy. All your robes are fragrant with myrrh and aloes and cassia" (Psalms 45:7–8); and, "My lover is to me a sachet of myrrh resting between my breasts" (Song of Songs 1:13). It was used by virtually all of the ancient peoples of Asia Minor in anointing and embalming the dead, including the celebrated morticians of ancient Egypt. In his 1770 treatise, John Cook offers an anecdote on the preservative properties of myrrh:

> *A bird, or any other small animal, or an insect, to be dipped several times successfully [sic] in the tincture of Myrrh it would soon be perfectly penetrated, or embalmed thereby, and converted into a kind of Egyptian mummy, capable of remaining entire for numerous ages.*

Copal

It is commonly assumed that hardened resin turns into amber at a specific age. Actually, the process is a continuum, from freshly hardened resins to those that are truly fossilized, and no single feature identifies at what age along that continuum the substance becomes amber. Generally, material that is several million years old and older is sufficiently cross-linked and polymerized to be classified as amber. Material that is only, say, several thousand years old is often referred to as *copal,* or subfossil resin. Copals are so incompletely cross-linked that a drop of alcohol or other solvent makes the surface tacky. Put close to a hot flame, copal will readily melt; amber will soften and blacken but not liquefy. The oldest copal deposit, from Mizunami, Japan, is approximately 33,000 years old. As expected, Mizunami copal displays characteristics between those of amber millions of years old and copal merely hundreds of years old. Copals will take a high polish, but since they retain volatiles from the original resin that readily evaporate, after a few years the surface becomes deeply crazed, like a dried lake bed. The extent of crazing depends on exposure to heat and air. Amber, too, will craze, but not as quickly or deeply as copal. Copal crazes so deeply, in fact, that this is a reliable way to distinguish material in old collections that is called amber but in fact is copal.

Confusion surrounds the use of the term *copal,* since some scientists also use it to refer to fossil resins of certain botanical origins. The major deposits of subfossil resins, or copal, were formed by tropical legume trees and the araucarians (any of a genus of conifer trees indigenous to South America and Australia), which are the true "copal trees" of chemists. Resins from these trees harden rapidly upon exposure to air, are distinctively hard, and have a higher melting point than other resins (but not more than amber). Yet another term, *resinite,* which is much more general and in use primarily by geologists, refers to any hardened resin, whether amber or copal.

Most copal occurs in the tropics or very wet temperate areas, generally where the trees that formed the copal still live. Since the tree species are extant, the source of the copals is quite certain. The most famous deposits are those that have been commercially exploited in the past for varnishes (now almost entirely replaced by synthetic resins), on the North Island of New Zealand and in East Africa. Copals from these regions were also the source of numerous forgeries in "amber."

On the North Island of New Zealand live huge kauri trees, the "sequoias" of New Zealand: *Agathis australis* and *Dammara australis.* Masses of resin from these trees ooze on and under bark (called *kauri gum*) and accumulate on the forest floor. Buried by hundreds or thousands of years of fallen needles, twigs, and branches, the subterranean kauri gum is sometimes found where the kauri forests no longer exist. At the peak of the kauri-gum industry, prior to the turn

A huge Agathis *tree in New Zealand, photographed in 1936. Trees such as this were the source of kauri gum.*

of the century, trees would even be tapped, although this was discouraged in order to protect the behemoths. Thousands of itinerant "gumdiggers" traveled among the various "gumfields." Most of them were Austrian immigrants, some of them poorer New Zealanders, and an occasional Maori. Export began about 1850; in 1856, approximately 1,440 tons were exported, and by 1906, exports reached 275,319 tons. Lumps of kauri gum ten to twelve pounds were not uncommon, and the largest one reported weighed nearly one hundred pounds.

Most copals derive from legume trees in the Caesalpinioidea group of families, especially the genus *Hymenaea*. A related genus of trees, with the appropriate name *Copaifera,* is the source of copals from Ghana, Guinea, and Sierra Leone in western Africa. *Hymenaea* copals occur in Minais Gerais, Brazil; eastern Dominican Republic; Colombia; and East Africa. Deposits from Santander, Colombia, are harvested for some especially large pieces (others in Colombia occur near Medellín and along the Magdalena River in Mariquita Province). Many of these impressive pieces contain termite swarms and other insect inclusions and are sold to amateur collectors as "Pliocene amber" (about two million years old), even though carbon-14 dating indicates it is only several hundred years old, like all the other *Hymenaea* copal deposits. Similarly, a clear *Hymenaea* copal from eastern Dominican Republic is sold as Dominican amber; true Dominican amber comes from the northern mountains and is light yellow to deep red. When the peoples of Asia Minor were tapping frankincense and myrrh trees for incense, copal and freshly hardened *Hymenaea* resin were burned as an incense by native peoples of Central and South America. The Maya burned it in special incense pots, and the Yanomamos of southern Venezuela still collect the resin for incense.

The only African species of *Hymenaea, H. verrucosa* (previously given its own genus, *Trachylobium*), occurs from Somalia to Tanzania, Zanzibar Island, Madagascar, and the islands of the Seychelles and Mauritius, some 1,000 miles off the East African coast. Around the turn of the twentieth century, *H. verrucosa* copal was the basis for a very lucrative industry: In 1898, some 512,600 pounds were exported to Germany for high-grade varnishes. Fresh pieces of the copal are a very pale, clear yellow, just like the New World *Hymenaea* copals. The American Museum of Natural History has a large collection of copal from Zanzibar, rich with insect inclusions. It has been suggested that some of the East African copal may be up to two million years old, but this is very unlikely.

Opposite: Large piece of copal from Santander, Colombia, containing beetles. Two surfaces are polished flat. Height 4.5″. American Museum of Natural History (Entomology)

Section of a copal tree (Hymenaea verrucosa) *from Zanzibar, off the coast of Tanzania. Between the bark and the heartwood is almost pure resin. The heartwood contains beetle galleries impregnated with resin. Diameter 4.8″. American Museum of Natural History (Entomology)*

Tertiary Amber

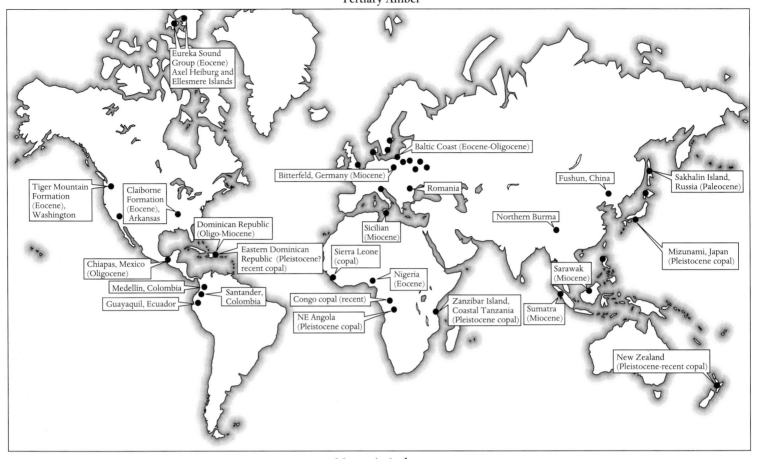

Mesozoic Amber

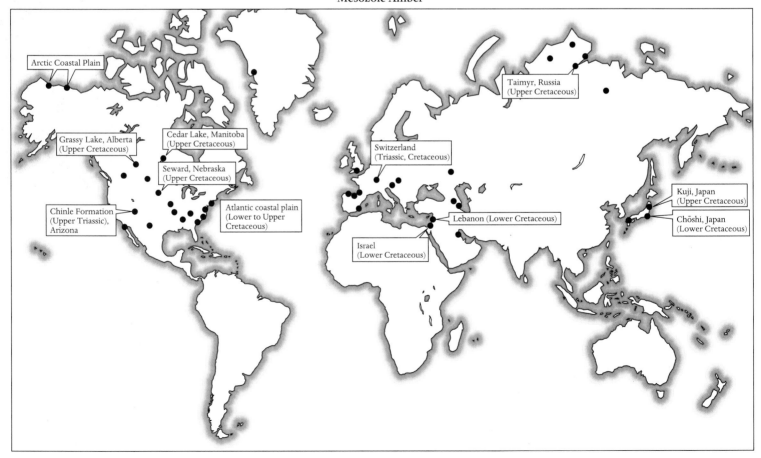

DEPOSITS OF THE WORLD

The Era of the Dinosaurs: Mesozoic Amber

The oldest "amber," perhaps more appropriately called a fossil resin, was hardly recognized as such when it was first discovered. Occasionally found lining some fine vessels from the trunks of *Myeloxylon* or other carboniferous pteridosperms (tree or seed ferns) are microscopically fine, black hairlike fibers that are actually resins some 320 million years old, although they are physically and chemically unlike any other fossil resin known, even the ambers. The second-oldest fossil resin exists in microscopic quantities in 260-million-year-old Permian limestone near the Chekarda River, in the western piedmont of the Ural Mountains.

From the Triassic period of Europe and North America derive dark red, highly brittle ambers, formed perhaps from extinct cycads like *Pterophyllum*. These Triassic resins, although considered true ambers, are also chemically unlike the younger ambers formed from conifers and flowering trees. One of the Triassic deposits is from Niederösterreich, Austria, about 60 miles southwest of Vienna. Another, the Raibler Sandstone Formation in Schliersee, Bavaria, is about 220 to 230 million years old. By this time, large vertebrates and most of the modern orders of flying insects had appeared. Not surprisingly, microscopic remains of organisms are found in the Schliersee amber, but they are of primitive organisms, such as bacteria, protozoa, fungal spores, and unidentified plant spores. An insect in amber this old would be sensational.

One of the biologically most interesting chapters in the history of Earth is the Cretaceous period, 140 to 65 million years ago. It was at the end of the Cretaceous that the dinosaurs died out. More important, it was during the Cretaceous that there occurred explosive radiations of the flowering plants, or angiosperms, and many modern families of insects. Today, the insects and flowering plants are supreme rulers on land: they comprise three-quarters of all life forms, as well as more biomass and more anatomical and chemical novelties than all other organisms combined. Without them, the world would be unrecognizable. A fundamental belief among biologists has been that the evolution of flowering plants and insects closely affected each other. Although there is some recent skepticism (that modern groups of insects evolved before flowering plants), most evidence indicates that the appearance of at least certain kinds of insects—like some beetles, flies, and certainly the bees, moths, and

*Fossil cone from the Upper Jurassic or Lower Cretaceous (c. 140 million years old), studded with amber. It was found in the Karzhantav Range, Chimkent region of southern Kazakhstan. The original cone must have been filled with resin. Length of cone .9".
Paleontological Institute, Moscow*

Opposite: Major amber deposits of the world

Opposite: Plate from O. Warburg, Beiträge zur Kenntniss der Vegetation des Süd und Ostasiatischen Monsungebietes, *1900, showing leaves, cones, and cone scales of the conifer* Agathis. *Araucarian trees such as* Agathis *are considered important sources of the ambers from the Mesozoic Era.*

butterflies—was linked with the evolution of angiosperm plants, and vice versa, and that this took place during the Cretaceous period. Fossils in Cretaceous amber have been a particularly revealing window for understanding this relationship.

Even though angiosperms were diversifying during the Cretaceous, the landscape at this time was probably dominated by cycads and conifers. All of the Cretaceous ambers are certainly coniferous, for arborescent (treelike) angiosperms probably had not evolved until the late Cretaceous; prior to this, they were herbs and bushes. In fact, for most of the Cretaceous deposits that have been chemically studied, the amber is thought to have been formed by an araucarian or araucarianlike tree. The Araucariaceae is one of six families in the Coniferae (three other, large families being the Pinaceae [pines, larches, spruces, and hemlocks], Cupressaceae [cedars, cypresses, junipers], and Taxodiaceae [sequoias and bald cypresses]). There are only two genera of araucarians living today, comprising thirty-one species, three of the species being among the ten tallest trees in the world (between 70 and 90 feet tall). The family is relict, confined now to portions of the Southern Hemisphere. Fossils of *Araucaria* from the Jurassic to the Tertiary, however, are scattered around the world. The family is a good candidate for the botanical origins of many Cretaceous ambers since, today, araucarians produce copious amounts of resin that becomes strongly hardened soon after exposure to air, which preserves it well.

J. Gürke ad nat. delin. et lith.

Meisenbach Riffarth & C.º Berlin.

A. Agathis macrostachys Warb., *B.* A. regia Warb., *C.* A. rhomboidalis Warb.,
D. A. borneensis Warb., *E.* A. philippinensis Warb., *F.* A. Beccarii Warb.

Verlag v. Wilhelm Engelmann, Leipzig.

In the absence of more information on conifer resins, however, it should not be taken for granted that any Cretaceous deposit is definitely of araucarian origin. For example, the 75-million-year-old amber from the Fruitland Formation in the San Juan Basin, New Mexico, is one of the rare cases where the amber has a definitive origin, since it is found among and embedded in logs and stumps of the amber tree, and it is Taxodiaceae.

There is a woeful lack of plant megafossils accompanying most amber deposits, including Cretaceous ones. Inclusions of plant fragments in amber can provide circumstantial or confirming evidence of the tree of its origin, but such inclusions are much rarer in Cretaceous than they are in Tertiary ambers. For example, there are no cones or conifer flowers known for any Cretaceous amber, although wood fragments and portions of needles and twigs occur in some.

All of the Cretaceous ambers are very brittle and highly fractured. Special preparation techniques are usually required in order to grind and polish a piece for a certain view of an inclusion, or even to keep the piece from disintegrating. Cretaceous amber is notorious for becoming crumbly when exposed to the atmosphere for several years. It is best preserved and prepared by embedding it in a synthetic resin.

A forest in New Zealand. With the exception of the Maori native in the forest, this is what a Cretaceous forest might have looked like some 90 million years ago, with tree ferns, cycads, and araucarian trees.

Although Cretaceous amber is found near Vienna and in Salzburg, Austria, it is the 100-million-year-old amber of France that is better known and probably more abundant. Occurring in the Paris and Aquitanian basins of northwestern France, near Bezonnais, Durtal, and Fouras, it resembles in both composition and kinds of inclusions the 90-to-94-million-year-old amber from certain deposits in New Jersey. Cloudiness of the amber is due to microscopic bubbles, and pyrite ("fool's gold") has intruded into cracks and even some of the insect inclusions. The pyrite has allowed high-resolution X-raying of some insects, since it is much denser than the surrounding amber and has replaced the original insect in faithful detail. As in most Cretaceous ambers, the fossilized insects within are tiny, less than one-tenth of an inch long on average, although some of the most interesting ones, like the termites and lacewings, are quite large. The insects are not plentiful. In the French amber, for example, one pound of raw amber yields approximately twenty insects and insect parts. Cretaceous amber from Canada, by contrast, yields about twice this number of inclusions.

North America
The most abundant sources of Cretaceous amber in North America are Alaska, several localities in Canada, and New Jersey. In 1955, a group from the University of California, Berkeley, collected amber from the shores of the Kuk, Omalik, and several other rivers on the northern Alaskan Coastal Plain. The paleontological value of the Alaskan material is limited because it is stranded on river shores; as a result, the pieces are small and heavily weathered, and a specific age is indeterminate.

The largest deposits of amber from Canada have yielded an exciting array of insects and other inclusions. The first deposit to be intensively studied, as early as 1891, was at Cedar Lake, Manitoba. The amber was so abundant (for a Cretaceous deposit) that in the early 1900s nearly a ton was collected for varnish. In the 1930s, Frank Carpenter, the great paleoentomologist from Harvard, collected several hundred pounds at Cedar Lake. Another large collection of the amber, much more closely studied than Harvard's, is in Ottawa. Since those collections were made, Cedar Lake has been dammed, inundating the deposits.

It has always been suspected that the Cedar Lake amber was redeposited from a distant source. Amber deposits from Medicine Hat, Alberta, in the Foremost Formation (about 75 to 78 million years ago) are chemically very similar to the Cedar Lake amber. Though presumably not the source of the Cedar Lake amber, the Medicine Hat amber may be contemporaneous and of the same botanical origin. Most recently, rich deposits of amber in Grassy Lake, Alberta, have yielded numerous tiny fossils, including a bird feather in one piece. Aphids are the most common insects in this amber, and among the most interesting are a pseudoscorpion, a praying mantis, and the oldest known mosquito.

The United States has several Cretaceous deposits, although only in New Jersey is amber found in appreciable quantities. The first insect discovered from

North American amber, a caddis fly, *Dolophilus praemissus* Cockerell, was identified in 1916 in amber from Coffee Sand, Tennessee; ironically, no recent collections have been made of this amber. A significant amber deposit also occurs in the Black Creek Formation (about 75 million years old) near Goldsboro, North Carolina. Amber from near Paden (Tishomingo County), Mississippi, in the Upper McShan Formation of the Eutaw Group (about 90 million years old), is found in small pieces sometimes up to 1.5 inches long. This amber is associated with fossilized wood of Cupressaceae, Pinaceae, and Taxodiaceae, so it is probably not derived from an araucarian tree. The amber is yellow to dark brown, mostly cloudy, and has been found to contain a host of fungal spores and hyphae but no insect as of yet.

Amber from the Atlantic Coastal Plain of the eastern United States has been known for more than 150 years, the first report of it being in 1821. That report described a piece found in clay near the shore of Cape Sable, Maryland, containing what was believed to be a gall made by scale insects. Amber has also been found on Cape Cod, Long Island, and Staten Island. Historically, large deposits on Staten Island were discovered in open pits mined for clay in brick

Opposite: Crude amber embedded in marcasite, a form of pyrite, from New Jersey. Length of largest piece 5.1".
American Museum of Natural History (Entomology)

Above: Variations in New Jersey amber: a large drop-shaped piece; opaque and oxidized pieces (center of top row and left column); and various transparent pieces. Length of largest piece 3.6".
American Museum of Natural History (Entomology)

Wood found with amber from New Jersey, presumably of the amber trees themselves. Length of longest piece 10". American Museum of Natural History (Entomology)

manufacture. These huge pits are now eroded in and covered with woodland. The amber was purportedly so plentiful that workers would pile it in barrels during the winter and burn it to keep warm!

Amber occurs in similar abandoned clay pits in Cretaceous exposures of New Jersey, where the most abundant North American deposits are found. Chemical analysis identifies the botanical source of the amber as araucarian; however, twigs in the amber and the microscopic structure of lignite found with the amber (sometimes the amber is found in the fossilized wood) indicate Cupressaceae. Amber deposits vary from 65 million to nearly 95 million years old, although an unusual Tertiary fossil "resin," with a consistency like solid, hard plastic, is known from New Jersey. Derived from the witch-hazel and sweet-gum family (Hamamelidaceae), it is composed of a material like polystyrene and is similar to *siegburgite,* known from Europe since the late nineteenth century.

The true, resinous Cretaceous amber from New Jersey is clear red to yellow. In some deposits, 70 percent of the material is turbid, being clouded with small bubbles and particles of debris swept off the bark as the resin streamed down the tree trunk.

Amber and other fossils from the New Jersey Cretaceous are revealing new insights into the flowering, literally, of the Cretaceous period. Two of the most important insect fossils are known from the New Jersey amber. One is a very primitive ant, *Sphecomyrma freyi* ("Frey's wasp ant"), described in 1966 by Harvard entomologists. Although older fossil ants have been described since then, this is still the oldest definitive ant. The ants are a successful group today, with about 14,000 living species, and they are pivotal components of some ecosystems, such as tropical rain forests.

The other fossil is a bee, *Trigona prisca,* which, incredibly, belongs to an evolutionarily recent group, the stingless bees, or meliponines. This bee fossil was unearthed in an old Museum collection; chemical analyses confirmed it was authentic New Jersey amber, as the label indicated (specifically, from Kinkora, New Jersey). Since the piece was not precisely documented, however, a dating more exact than 65 to 80 million years old has not been possible. Nonetheless, such an anomalously advanced insect of this age raised controversy, especially since it had serious implications for the evolution of flowering plants. Since all bees forage on pollen and nectar, such an advanced bee in the Cretaceous would indicate a corresponding antiquity for the angiosperms. Despite the chemical analyses, some scientists remain skeptical about the authenticity of the specimen.

In the 1990s, new excavations in New Jersey, from the same Cretaceous deposits that yield much of the amber, have revealed a stunning array of 90-million-year-old flowers. They are preserved, not in amber, but in clay. The flowers are tiny and made entirely of carbon, probably transformed this way when forest fires "charcoalified" what lay beneath the forest leaf litter. Preservation is perfect: stamens, anthers, pollen, stigmas, petals, glands, and the cells that make them up are all visible. In many cases these flowers are better preserved

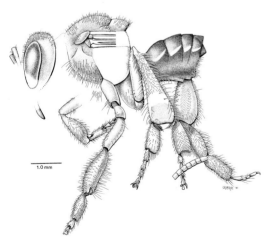

1.0 mm

The oldest known bee, Trigona prisca, *fossilized in New Jersey amber. American Museum of Natural History (Entomology)*

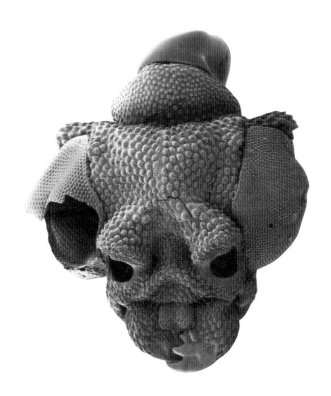

These tiny carbonized flowers and a beetle head (lengths of each about .1″), preserved in clays with the New Jersey amber, were photographed with a scanning electron microscope. L. H. Bailey Hortorium, Cornell University

Left: Ericalean flower shown intact (above) and "dissected" (below). It has petals, round sepal glands along the petals, and nectaries at the bases of the stamens, which were presumably used for attracting insect pollinators, such as bees.

Right, above: Detrusandra, a relative of magnolias, is much simpler and was perhaps wind pollinated.

Right, below: Head of a cupedoid beetle revealing intricate sculpturing. This mode of fossilization is the closest equivalent to preservation in amber. These and similar fossils complement those preserved in the New Jersey amber, which is of similar age.

than examples in Tertiary ambers. Many of the flowers are from plants surprisingly advanced evolutionarily, belonging to tropical families and other groups, which may explain such an advanced bee in New Jersey amber. For example, there are flowers of laurels (Lauraceae), Chloranthaceae, tiny magnolia-like flowers, and from plants related to the heaths (family Ericaceae) and the witch-hazel family, Hamamelidaceae. Pollen in the ericalean flowers was held together in clumps by threads of a viscous substance. This is strictly a feature of flowers that are pollinated by insects, which serve to make the clumps of pollen adhere to the hairs of an insect like *Trigona*. Others have glands that secrete scents to attract insects. The hamamelidaceous fossil flowers have nectaries near the petals, which are other hallmarks of flowers pollinated by insects.

Besides a variety of small organisms in the amber, the new excavations of amber in New Jersey have found other insects that belong to living genera. At least some species of insects had close relatives that extended back nearly 100 million years. We are gradually learning that the New Jersey bee is not anomalously old: the Cretaceous is anomalously young.

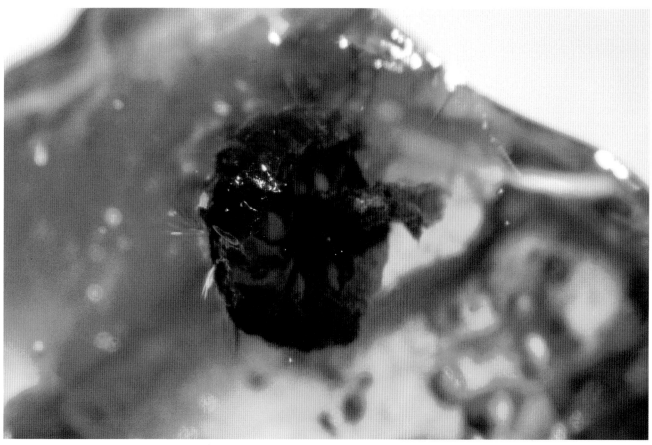

Japan — Among seashore deposits in Kuji, Japan, can be found 85-million-year-old amber, from the Taneichi and Kunitan Formations. The amber occurs with marine fossils like mosasaur teeth and ammonites (extinct relatives of the nautilus) and comes in a remarkable variety of colors and opacity, much of it an appealing caramel color. It is the oldest amber in the world from which objects have been carved, one reason being the large size of some pieces. One piece, found in 1927 near Kuji, weighs 44 pounds; another, found in 1941 (and now in the National Science Museum in Tokyo), weighs 35 pounds. Both of these large pieces are opaque yellowish orange. Even older amber (about 120 million years old) has been found in Chōshi, Japan. All the Cretaceous ambers have some insects; one piece from Kuji has portions of a bird feather.

Variations in the 85-million-year-old Kuji amber from Japan. Length of largest piece 3". American Museum of Natural History (Entomology), courtesy of Kuji Amber Museum

Opposite: Collecting late Cretaceous amber from Kuji, Japan

Collecting Cretaceous amber from the Taimyr Peninsula, northern Siberia

Above: En route to Romanikha, eastern Taimyr. Larches here are the northern-most forests.

Middle: Baikura-neru Bay, on the edge of Lake Taimyr in the center of the peninsula. This site has yielded most of the fossiliferous amber.

Below: Yantardakh (Amber Mountain), eastern Taimyr. Digging amberiferous lignite out of a "lens" embedded in a wall of sand and clay

Opposite: Screening and washing the amber at Yantardakh. The person on the left is examining pieces for inclusions with a hand lens.

Siberia Probably the largest deposit of Cretaceous amber in the world comes from the Taimyr Peninsula in northern Russia. The oldest report of this amber was made as early as 1730. Of the four main deposits on or near the Taimyr Peninsula, one, about 80 million years old, is from the Khatanga Depression, also the site of the northernmost forests (larches). In both western and central Taimyr are 100-million-year-old deposits from the Cenomanian-epoch, Dolganian and Begichev Formations. Another is from the Arctic Institute Island, just off the west coast of Taimyr. Scientists at the Paleontological Institute in Moscow have spent decades excavating and screening this amber for the countless tiny organisms fossilized in it.

The Middle East The oldest amber in the world containing insects and other larger organisms comes from the Middle East, specifically Lebanon, although similar amber occurs in Israel and Jordan. The amber is chemically similar in all of these areas and is from the Neocomian age (Lower Cretaceous, about 120 to 130 million years old). The largest amounts of amber are found at Dahr al-Baidha, between Beirut and Damascus, and around Jezzine. Only two collections of Lebanese amber exist, one at the Museum für Naturkunde, Stuttgart, the other being the Acra Collection, part of which is at the American Museum of Natural History.

Screening and preparing inclusions in Cretaceous ambers is extremely tedious because of the many fractures. The Acras spent several decades processing approximately 200 pounds of raw amber and accumulated a wonderful collection of more than a thousand fossiliferous pieces. In that collection are many exciting earliest geological records of various arthropods, such as termites and the oldest definitive moths (including a caterpillar).

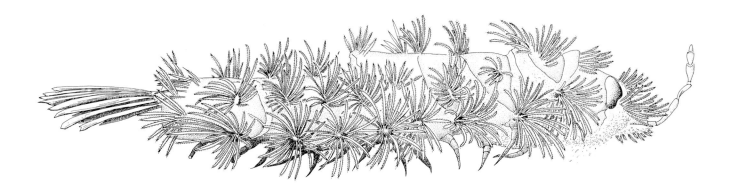

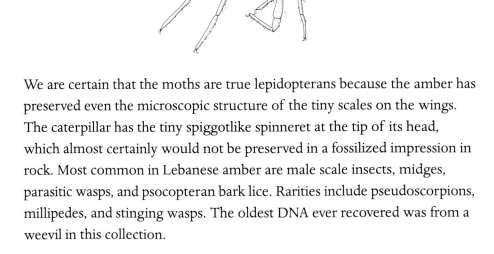

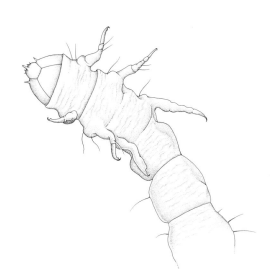

We are certain that the moths are true lepidopterans because the amber has preserved even the microscopic structure of the tiny scales on the wings. The caterpillar has the tiny spiggotlike spinneret at the tip of its head, which almost certainly would not be preserved in a fossilized impression in rock. Most common in Lebanese amber are male scale insects, midges, parasitic wasps, and psocopteran bark lice. Rarities include pseudoscorpions, millipedes, and stinging wasps. The oldest DNA ever recovered was from a weevil in this collection.

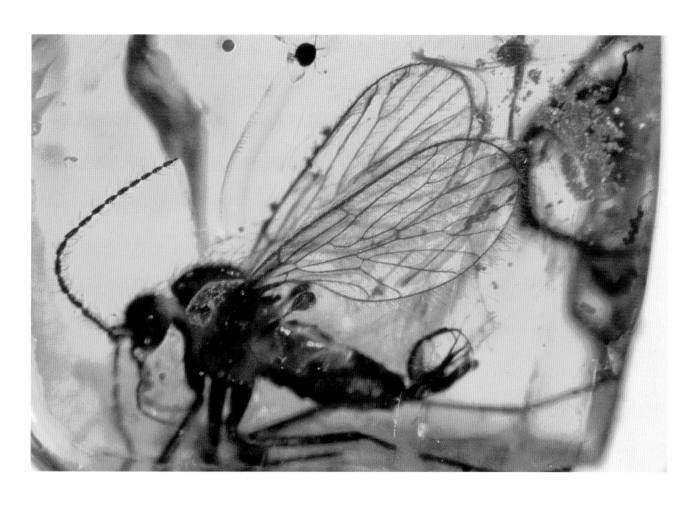

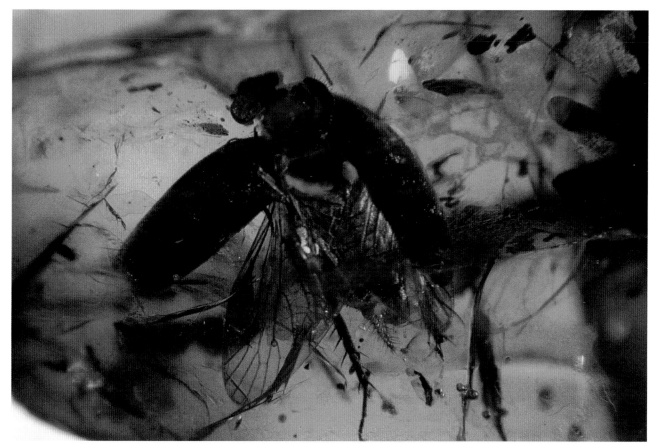

Tertiary Deposits

Among the dozens of major amber deposits scattered around the world, most are from the Tertiary period, which extends from 65 million years ago to the present. Even within this period, most deposits derive from the Eocene, a few from the Oligocene and Miocene ages, and even fewer from the other ages in the Tertiary. The botanical sources, colors, and composition of these ambers are extremely varied, unlike the earlier Cretaceous ambers, which are mostly a brittle, transparent yellow to red (perhaps reflecting more uniform botanical origins).

Above: Excavating the largest piece of amber in the world, in Sarawak. It is now on display at the Museum für Naturkunde, Stuttgart.

Chunk of amber from Sarawak. Very opaque and blackish, it is from the same locality as the piece of amber above. Height 4.4". American Museum of Natural History (courtesy of Museum für Naturkunde, Stuttgart)

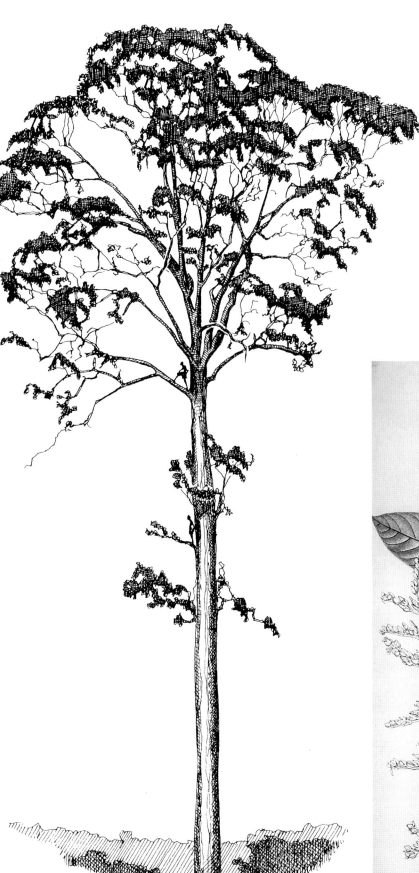

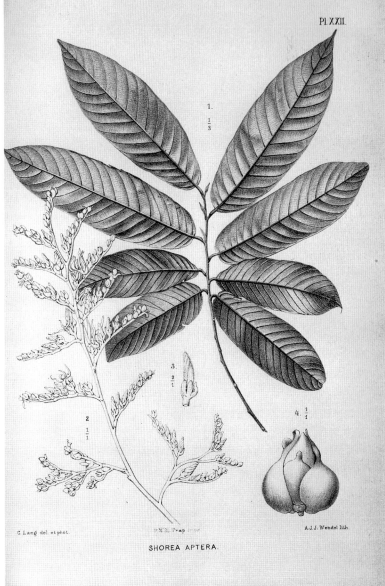

Left: Giant Shorea curtisii *tree on Brunei. Note two men in the tree, about midway up on the left. Burmese legend holds that Gaudama died and perhaps was even born in a grove of Shoreas:*

Leaves, flowers, and fruit of Shorea aptera. Extinct species of Shorea *or some other dipterocarp tree apparently gave rise to the amber from Sarawak in Malaysia and Arkansas in the United States.*

Pl. XXII.

1. $\frac{1}{3}$

3. $\frac{2}{1}$

2. $\frac{1}{1}$

4. $\frac{1}{1}$

C. Lang del. et phot.

P.W.M. Trap impr.

A.J.J. Wendel lith.

SHOREA APTERA.

The largest piece of amber in the world, deriving from the lower to mid-Miocene Nyalau Formation of Sarawak, Malaysia, was discovered on December 3, 1991. It weighs more than 150 pounds and, to transport it to the Museum für Naturkunde in Stuttgart, Germany, where it is now on display, it had to be sawed into several sections. The amber itself is similar to dense coal, impregnated with the fossil resin. Upon polishing, various colors of the Sarawak amber become apparent: white, pink, orange, green, even violet, although a clear yellow is rare. Microscopic reddish-brown droplets impart the pinkish opacity. The Merit-Perla area, where the piece was recovered, is mined for coal, and amber is found among some of the coal seams. Bright, yellowish amber occurs in some seams, in pieces 1 to 40 centimeters in diameter. The amber found so far has contained centipedes, spiders, beetles, ants and wasps, and various flies. The Dipterocarpaceae family is the apparent tree source of this amber. Many species of dipterocarps grow in Asia, where they are valuable timber trees because of their girth, straight trunks, and resinous wood, which helps prevent termite infestation.

Amber mines in northern Burma, c. 1930

The largest piece of transparent amber, which is very deep red, from Burma (Myanmar). It weighs 33.5 pounds and is 19.5" long. Natural History Museum, London

Historically, the best known Asian amber is *burmite,* from the Hukawng valley of northern Burma (now Myanmar). It was reported in European scientific literature as early as 1836, although mining had been done here for at least a millennium. By the 1930s, many of the amber mines, at least near Maingkwan and the village of Shingban in the Hukawng valley, consisted of hundreds of abandoned pits overgrown by dense jungle. At that time, the largest workings were at Khanjamaw, where 150 Kachins, Shans, and Shan-Chinese were digging 500 to 600 pits. Most of the pits were 30 to 40 feet deep, occasionally 50 feet, depth being constrained by the appearance of a deep sand layer and water, which seeps in at 40 feet. To keep pit walls from collapsing, elaborate screens of bamboo poles supported by wooden posts were needed.

Today, burmite has almost legendary appeal, in part because the deposits are no longer mined and the supply is generally unavailable. (This may be because the mines are exhausted; information on Chinese amber mining is sketchy.) The appeal is also due to burmite's properties. From the few scientific collections of it existing (the best being at the Natural History Museum, London), we know that it was highly fossiliferous. Fourteen local varieties were recognized, most of them a rich, transparent red with strong ultraviolet fluorescence. Burmite is also harder than most other ambers and is excellent for carving. In fact, much of it was exported to Yunnan, in southern China, then to Beijing, for carving various objets d'art. The largest specimen of transparent amber in the world is a deep-red piece of burmite weighing 33.5 pounds, in the mineralogy department of the Natural History Museum, London. Discovered in 1860, it was presented to the museum in 1940.

Interestingly, there is no historical mention of amber from Liaoning Province, China, for use in carving ornamental objects. Amber here occurs with coal in the Guchenzgi Formation of Fu Shun. It too exists in large pieces and is highly fossiliferous. Burmite and Fu Shun amber were both formed in the Eocene, and the botanical sources are unknown.

Africa Africa is well known for its production of copal, but the only Tertiary deposit on the continent with true amber comes from southeastern Nigeria near Umuahia, in the Ameki Formation of the Eocene. The amber is dark red, transparent to opaque. No biological inclusions are known, nor is the plant source.

Europe Despite the overwhelming size of the Baltic deposits, Sicilian amber has its own allure. Amber from the Simeto River of Sicily, near Catania, and the Salso River (called *simetite*) is renowned for its varied, deep colors: red, blue, and smoky green. Arnold Buffum, who extolled the virtues of Sicilian amber in his book *The Tears of the Heliades,* amassed a wonderful collection of European amber objets d'art in the late nineteenth century, which are in the Museum of Fine Arts, Boston. Other collections of simetite, but of mineralogical specimens, are at the American Museum of Natural History. Although some pieces are indeed the deep red that Buffum described, there are no pieces with distinctive green and blue hues in the Boston collection, perhaps because these colors have faded. Sicilian amber is younger (Oligocene to Miocene) than the Baltic amber, and the deposits are much smaller. Simetite is collected only rarely today, although it is hard to imagine that an exhausted supply is the reason, for there never was organized or mechanized mining of it like that done for amber on the Samland Peninsula of the Baltic Sea.

Obscure by European amber standards is *rumanite,* from the Carpathian Mountains of Romania. Cretaceous and Tertiary deposits of Romanian amber have been found. The true Tertiary rumanite has properties and colors similar to simetite and comes largely from areas surrounding Colti, in Buzau District. Elsewhere in the Carpathians, in the Lvov and Ivano-Frankovsk regions, near the town of Verkhnii Sinevidnyi, are Eocene deposits of succinite. Simetite and most rumanite lack succinic acid, and the botanical source of the ambers is uncertain.

North America   Although overshadowed by the vast deposits of the Baltic region, North American deposits from both the Cretaceous and the Tertiary periods are still quite varied. Some deposits are surprisingly rich. The northernmost amber in the world occurs in early Eocene deposits of Axel Heiburg and Ellesmere islands, in the Canadian Arctic. The amber formed in the permafrost and can include remarkably well-preserved tree stumps, cones, and other plant fossils, as well as the fossils of extinct catfish, snapping turtles, and plagiomerid mammals related to early primates, evidence of the subtropical

Cones and branches of the relict, living pine Pseudolarix kempferi. *Forty-million-year-old cones of an extinct* Pseudolarix *from the northernmost islands of Axel Heiburg and Ellesmere have amber with large amounts of succinic acid. Thus,* Pseudolarix *may be the kind of tree that gave rise to Baltic amber, whose botanical origins have been controversial.*

Right: Landscape on Axel Heiburg Island in the Arctic Circle, where 50-million-year-old forests are preserved in the permafrost with amber

Amber from Axel Heiburg Island, with a fossilized cone in the center, perhaps from the kind of tree that produced the amber. Most of the amber is heavily weathered. Length of cone .8".
American Museum of Natural History (Entomology)

Variety in the 40-million-year-old amber from central Arkansas. Some pieces have been polished, revealing transparency; others are completely opaque. Length of largest piece 1.7". American Museum of Natural History (Entomology)

remains beneath the frozen desolation. The amber itself is not so well preserved, much of it having an oxidized, powdery, deep crust, with a small core of transparent yellow amber. Amber from fossilized *Pseudolarix* trees found here contains succinic acid in amounts similar to that found in Baltic amber, which may relate these trees to the botanical sources of Baltic amber.

The largest North American deposit of amber is from the Eocene Claiborne Formation of Malvern County, Arkansas. The amber is in two locations, one an expansive pit mined for clay to manufacture bricks, the other an abandoned clay pit. In the active mine, pieces up to three inches long can be found on the surface of a dark clay impregnated with lignite. If the appropriate stratum is exposed, it is possible to collect several pounds off the surface in one day. This amber is very distinctive for its weathered rind and dense internal flow lines, which are also weathered. Intact amber in the core of a piece can be red to yellow; the yellow amber is more often made slightly cloudy by microscopic bubbles. Myriad arthropods are preserved in the amber, but finding them requires diligence because of the opacity of the material. Chemistry suggests that the botanical source is in the Dipterocarpaceae, which is intriguing since no trees in the family grow now in North America.

A much smaller but interesting deposit of Tertiary amber is found in the mid-Eocene Tiger Mountain Formation near Seattle, Washington, in a small, steep exposure in heavily wooded state property. The amber, a dark, transparent red, is extremely brittle and fractures easily when extracted from the clay substrate. Although no insects have been found in it, the many plant fibers in the amber are similar to those on the bark of cedars in the Cupressaceae, which suggests that the amber was formed from a tree in the same family.

Baltic Amber

The largest deposits of amber in the world, and the ones exploited the longest, derive from the shores of the Baltic Sea in northern Europe. Amber also washes up on the shores of eastern England and Scotland. Baltic amber has an exceptionally rich history of ancient trade, supported by guilds of amber craftsmen and stunning works of art. It even figures in Greek mythology.

Location and Geology The true Baltic amber is found on or near the shores of the eastern Baltic Sea, particularly on the Samland Peninsula. The peninsula, a mere 400 square miles in size, has produced 90 percent of all the amber in Europe. Both its northern bay (Kurisches Haff) and southern bay (Frisches Haff) are nearly entirely closed off to the Baltic Sea. Beaches on the side of the peninsula facing the sea are narrow, with steep, vertical cliffs. Amber washed up on the beaches, especially after storms, has been harvested for at least ten millennia. A few huge pieces have been found;

Samland Peninsula, Baltic coast

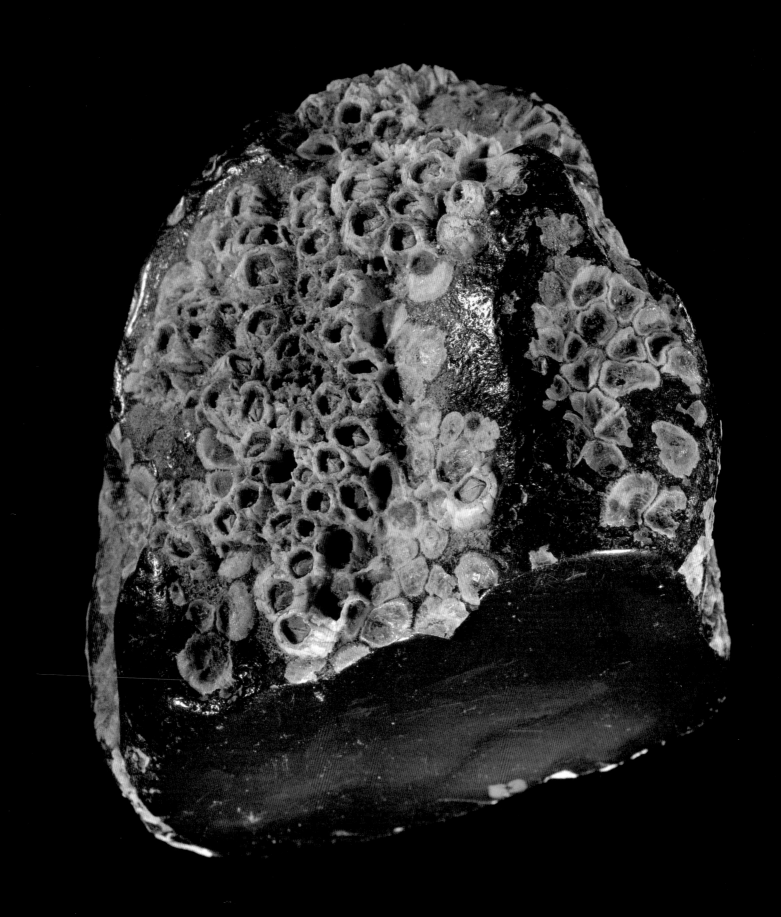

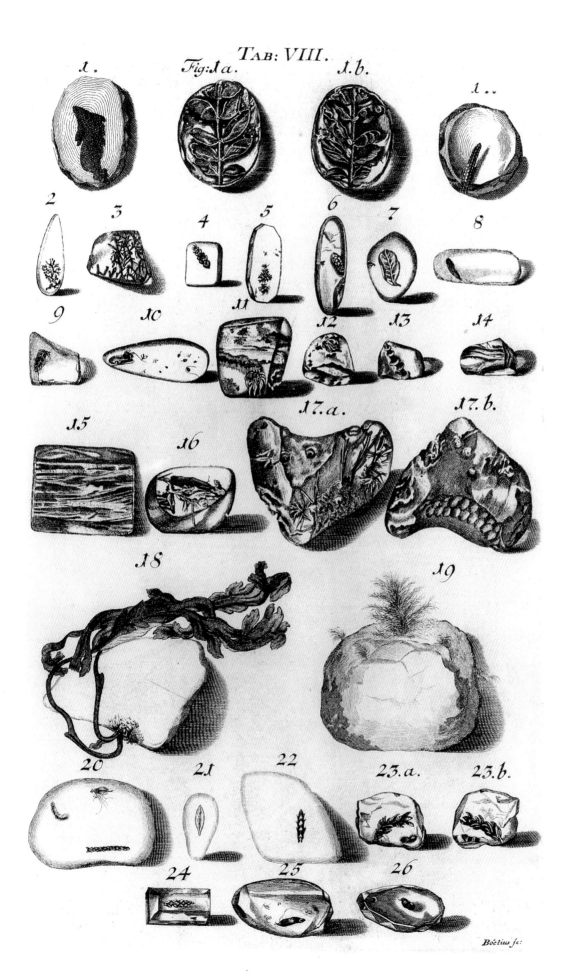

TAB: VIII.

Opposite: Section of Baltic amber encrusted with barnacles, with one end cut and polished. Pieces like this demonstrate that Baltic amber was in marine water after being eroded from sediment. Length 3.4". American Museum of Natural History (Earth and Planetary Sciences)

Plate from Nathanael Sendelio's 1742 monograph on Baltic amber, Historia Succinorum Corpora aliena involventium et Naturae Opere.... He depicts plant, wood, and insect inclusions, as well as several pieces with sea-weed. Marine organisms became attached to some pieces of amber that were deposited by seawater.

one, weighing 21.5 pounds (now in the Humboldt Museum, Berlin), was discovered in 1890 at the mouth of the Oder River. Amber is stranded on other Baltic shores as well and, occasionally, even the eastern shores of England. In North Jutland, Denmark, for instance, 3,000 pounds were collected this way in 1800, and, after several stormy years between 1822 and 1825, one Danish merchant collected 686 pounds at Ringkjöbing (one piece purportedly weighed 27 pounds). The record, though, goes to the Samland Peninsula; during one day in 1862, 4,400 pounds were collected off beaches in the town of Palmnicken (now Yantarny). It is not coincidental that the most productive amber mine in history was established in that town about ten years later.

Amber had been collected largely from the Samland beaches up until the mid-nineteenth century, when two people made a major impact on massive-scale mining of amber. In 1850, Königsberg's Society for Physical Economy hired geologist George Zaddach, who described how the amber was concentrated in layers of the *blau Erde* ("blue earth," actually greenish and formed by glauconite)

Large piece of Baltic amber, left unpolished to show the natural fissures, with a necklace of polished amber beads. Length 9.8". American Museum of Natural History (Earth and Planetary Sciences)

Such a large piece of amber would have been prized by artisans for sculpting a figure or other decorative piece.

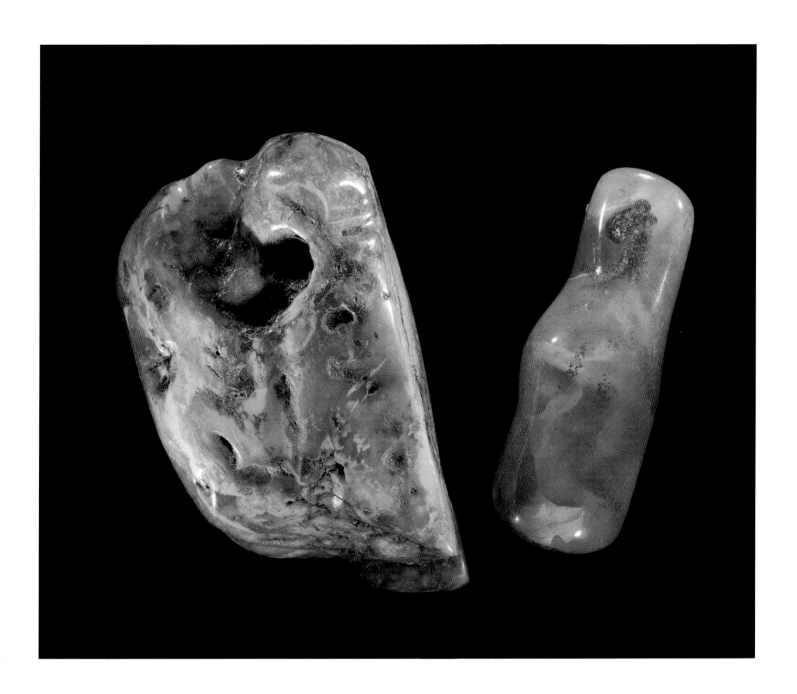

Two pieces of Baltic amber with
surfaces polished. One piece is a
mottled, opaque yellow-orange with
deep pits. The other is mostly
transparent with milky swirls on the
interior. Length of larger piece 4.4".
American Museum of Natural History
(Earth and Planetary Sciences)

The Palmnicken amber mine, c. 1985

dating back to the Eocene epoch, 40 million years ago. The *blau Erde* everywhere was 16 feet below sea level, and some 130 to 150 feet below the topsoil. More important, this layer was submerged on the floor of the Baltic Sea and reached the Samland Peninsula at only a few locations, one being Palmnicken. What was washed up on the shores was merely cast off from the bed of the sea; a much richer source remained to be tapped. In 1854, engineer Wilhelm Stantien from Memel began dredging operations for the amber, 35 feet down from the floors of the Haffs. By 1865, the mining firm of Stantien and Becker was operating twenty-two steam barges and employing about one thousand people. In 1868, they collected an unprecedented amount of amber: 185,000 pounds. By 1870, Stantien and Becker began open-pit mining, and the famous Palmnicken mine was opened in 1875. In its first year, Palmnicken generated 450,000 pounds of amber, and its yield improved steadily until 1895, when the unbelievable amount of 1.2 million pounds was extracted. One Felix Dahn described Palmnicken, where there worked "hundreds of men, women, and children, in all imaginable costumes, in the oddest of attires, shielding themselves against the sharp, whistling winds, digging vigorously and swinging their shovels to the languid strain of some sombre melody."

By 1930, amber extraction at Palmnicken was largely mechanized. Huge conveyers dumped buckets of *blau Erde* into open freight cars. The trains then carried the earth over to grates, where it was spilled to the spray house below and blasted with high-pressure hoses; small pieces floated out of a slurry, larger pieces were collected by hand. Out of the hundreds of thousands of pounds extracted yearly, nearly 90 percent was of poor quality and suitable only for chemical processing; the remainder was used for carvings and jewelry or contained fossilized inclusions. That 90 percent was dry distilled in huge iron retorts, which yielded 60 to 65 percent amber colophony (a high-grade varnish), 15 to 20 percent amber oil (used in medicines, casting, and the highest grade varnishes), and 2 percent distilled acids (used for medicines and dyes). Palmnicken is still the most prolific amber mine in the world.

<u>Features</u>   About 90 percent of Baltic amber has a high concentration of succinic acid (up to 8 percent), from which the name *succinite* is derived. Agricola (Georg Bauer) is said to have been the first to isolate succinic acid from Baltic amber, around 1546. Some Baltic amber, a yellow, friable amber called *Mürber Bernstein*, lacks succinic acid and has alternatively been named *Gedanite*. Other, rarer Baltic ambers, also lacking succinic acid, are *stantienite* and *beckerite*, both of which are opaque, dull brown, or black. *Glessite,* the rarest form, is yellow and softer than succinite. Even within succinite there exist various classes, distinguished by the size of numerous bubbles in the amber. *Foamy* amber is caused by a froth of larger bubbles, while *bone* amber is marked by microscopic bubbles. Bone amber is white to yellowish opaque, like ivory, and was eagerly sought for particular portions of carvings, such as inlays. Pieces having bubble sizes between those of foamy and bone amber are called *flom,* or goose-grease, and *bastard* amber. Bastard amber is clouded by milky swirls and is the most common of the opaque varieties. Why some ambers are opaque and others are not is not well understood. The fact that some Baltic amber lacks succinic acid suggests that several different kinds of trees may have given rise to the Baltic amber.

<u>Botanical Origins</u>   Exactly what tree or trees gave rise to the amber from the Baltic region has long been a matter of controversy and confusion (possibly resolved just recently), and study of botanical inclusions in Baltic amber has a distinguished history. In 1836, the German botanist H. R. Goeppert described the Baltic amber tree as *Pinites succinifer.* The tree was identified from microscopic features of wood fragments preserved in the amber, which, Goeppert believed, also showed similarities to pines. In fact, another botanist later assigned the Baltic amber tree to the genus of true pines, *Pinus.* Other evidence in favor of a pine or pinelike origin are the many cones and needles in the amber. Baltic amber, however, lacks abietic acid, which chemically distinguishes pine resin. The alternative hypothesis states that Baltic

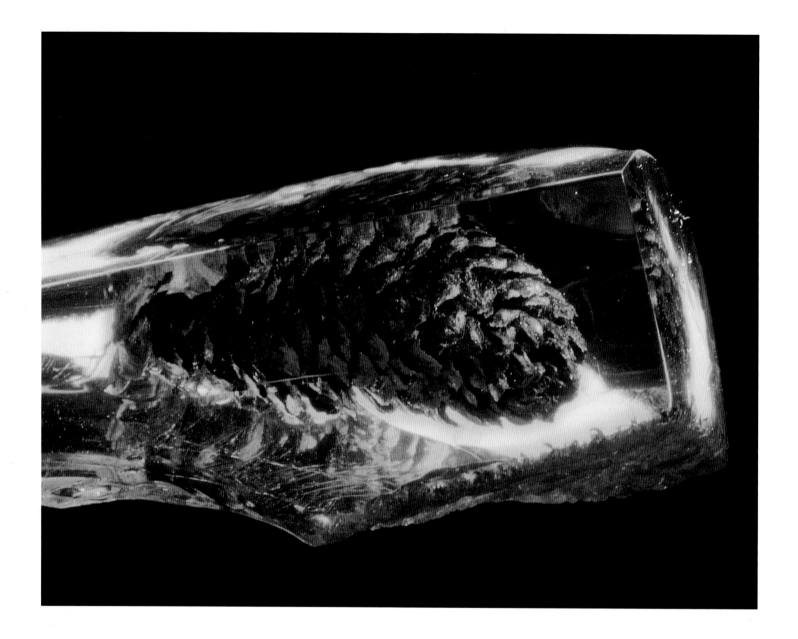

Opposite: A small flower (diameter
.6″) in a much larger piece of Baltic
amber. Private collection

amber was formed from an araucarian or a tree like one, but araucarian resin does not have the succinic acid that is so distinctive of most Baltic amber. In addition, there are few araucarian fossils in the Northern Hemisphere, and apparently none in Baltic amber.

A recent discovery that sheds considerable light on the origins of Baltic amber is that some living trees in the pine family, which belong to the genera *Keteleeria* and *Pseudolarix,* do indeed produce succinic acid. The latter is of particular interest since resin in 40-million-year-old *Pseudolarix* cones from Axel Heiburg Island in the Canadian Arctic also contains succinic acid. *Pseudolarix* today is found in Asia, and one species, *Pseudolarix amabilis,* is very narrowly restricted to some mountains in eastern China (the other two species have higher concentrations of succinic acid). The fact that 40-million-year-old trees that produced succinic acid existed on (what are now) the northernmost islands strongly suggests that *Pseudolarix* could have been in Scandinavia at the same time, and thence in the Baltic region. The *Pseudolarix* hypothesis is also bolstered by the fact that many of the other plant and insect species fossilized in the Baltic amber are closely related to species now living in Asia, Australia, and even Chile.

Inclusions Given the mountains of Baltic amber extracted from the Palmnicken mine alone, one can only imagine the thousands of pieces that were found containing interesting inclusions. The amber collection of Albertus Universität in Königsberg absorbed the amber collection of the Stantien and Becker firm, which in 1914 totaled some 70,000 pieces. Harvard's Museum of Comparative Zoology has a superb collection of 16,000 fossiliferous pieces, many brought over from Europe in 1867 by Hermann Hagen, a brilliant entomologist from Königsberg who became a professor at the Harvard museum. The Königsberg collection was by far the largest in existence and, by all accounts, was

A famous specimen: larva of the owl "fly," Neadelphus protae, in Baltic amber, in which all of the intricate processes and hairs are preserved. Like the related ant lions, these larvae impale their prey with their huge mandibles and then suck them dry. Hagen Collection, Museum of Comparative Zoology, Harvard University

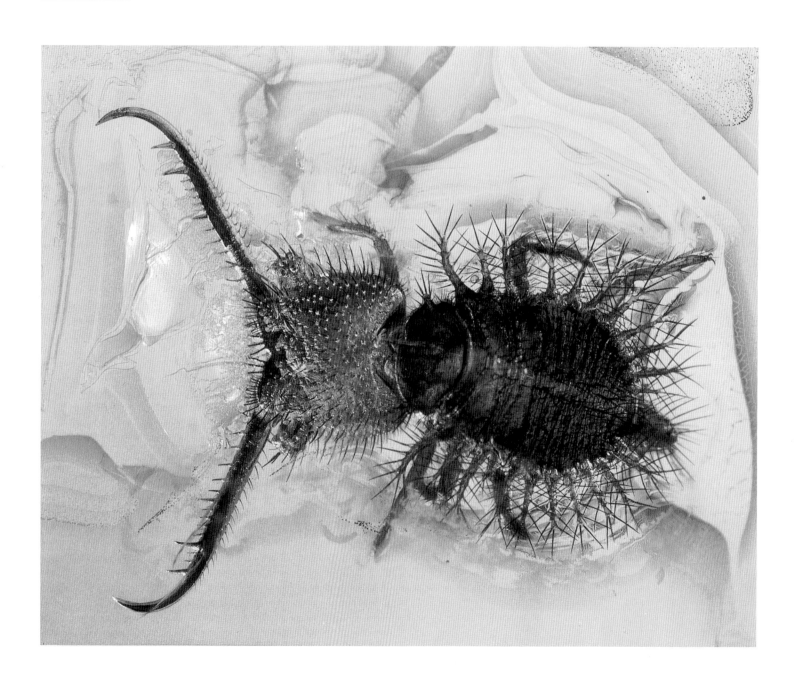

destroyed by fire in World War II. Actually, only portions of it were lost, because during the war the collection was divided among various localities for safekeeping, one place being the Institut für Paläontologie in Göttingen, where part of the collection still resides.

German scientists developed the paleontological study of amber fossils for several reasons. One, of course, was the proximity of the richest amber deposits in the world. The other was their perfection of optics, specifically in microscopes. Since 1800, hundreds of specialized scientific papers have described myriad organisms in the Baltic amber. Some of the nineteenth-century monographs, particularly the botanical ones, are illustrated in lavish detail with copperplate etchings, hand painted with watercolors. For G. C. Berendt's lovely 1830 monograph on the flora of the Baltic amber, he had studied more than 2,000 pieces with plant inclusions. Hugo Conwentz's botanical monographs of 1886 and 1890 are the most complete, with a stunning delicacy that modern scientific illustration could never hope to accomplish. Continuing the prestigious German tradition is Dieter Schlee at the Museum für Naturkunde, Stuttgart, home to the most comprehensive collection of ambers around the world and a huge collection of fossiliferous pieces, from the Baltic, Dominican Republic, and Lebanon.

Sven Larsson's 1978 book, *The Paleobiology of Baltic Amber,* summarizes nearly 150 years of scientific work on Baltic amber. Dating methods are far too imprecise to confirm if all Baltic amber is 40 million years old, but we know that succinite can be as young as 20 million years old, which is the age of the huge deposits of amber from coal mines in Bitterfeld, Germany (these mines are no longer active, but they did yield extensive collections of fossils, now at the Humboldt Museum in Berlin). If the fossilized organisms did exist all at the same time, the great diversity of tiny animals, plants, and fungi allows a very thorough reconstruction of the ancient amber forest. Such a reconstruction is presented later for the Dominican amber forest, which contains tropical species not unlike ones living on Hispaniola today (see page 101). Perhaps a result of being 10 millions years older, when sea levels and changing climates had more effect, the Baltic amber biota was distinctively subtropical.

The gamut of diversity in the Baltic amber includes bacteria, slime molds (actually, they are colonial protozoa), true molds, parasitic fungi, higher fungi (like mushrooms), lichens, mosses, ferns, cycads, conifer cones, flowers of nearly a hundred species of plants, and hundreds of species of arthropods. Stellate plant hairs (trichomes) are more common in the Baltic than in most other ambers. Since oak flowers occur in the amber, these trichomes are often attributed to oaks, even though trichomes are widespread throughout the flowering plants. Not surprisingly, many of the insects are forms whose living relatives are found on dead and decaying tree trunks and under bark. Swarms of insects infested injured and rotting wood, perhaps the result of "succinosis," Conwentz's hypothetical disease that led to the demise of the forest and the formation of such prodigious amounts of resin. Besides stellate plant hairs,

Piles of crude amber being bagged at the Bitterfeld amber mine. Now closed, it had yielded an exceptional amount of amber.

Opposite: A plate from Hugo Conwentz's 1890 monograph on Baltic amber flora, Monographie der baltischen Bernsteinbäume, showing flower inclusions and leaf impressions in the amber

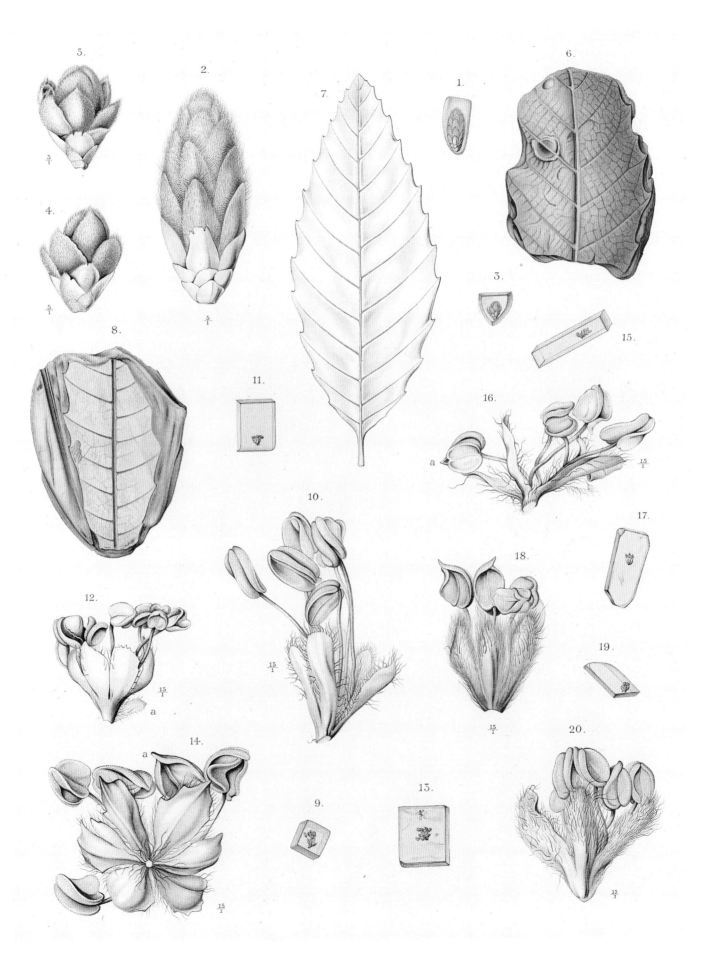

another distinctive feature of Baltic amber inclusions is the preservation. Insects in Baltic amber, much more than in any other amber, have a milky covering (*Schimmel*) over at least a part of the body. This milkiness is an emulsion of microscopic bubbles caused by decomposition.

Collectors of Baltic amber are not immune from the obsession for vertebrate remains typical of most amber deposits. Unfortunately, there is also a venerable history of forgeries in Europe (see pages 140–41). The only whole, possibly authentic vertebrate specimen in Baltic amber was a *Nucras succinea,* a small lizard related to ones living now in Africa, which apparently was lost with some of the Königsberg collection. Small tufts of mammal hair and a few small single feathers occur in Baltic amber, and recently the tails of a lizard and a rodent were found.

One of the most important insights into evolution that the study of Baltic amber fossils has made concerns extinctions. For many of the now extinct plants and insects in this amber, the closest living relatives are found in tropical or subtropical Asia, Australia, or southern South America. For example, the small parasitic plant *Trigonobalanus* today grows in Southeast Asia. The plant *Trianthera* in Baltic amber is closely related to *Eusideroxylon* from Borneo and Sumatra. Archaeid spiders and many chironomid midges have their closest living relatives in New Zealand, Australia, or Chile. Why some groups of organisms were once widespread and became extinct throughout most of their range is uncertain.

A plate from H. R. Goeppert and G. C. Berendt's 1845 monograph on the flora of the Baltic amber, Die Bernstein und die in ihm befindlichen Pflanzenreste der Vorwelt, *depicting details of cones and flowers, with the actual size of the specimen by comparison*

Dominican and Mexican Amber

Supplanting the popularity of Baltic amber, at least in North America, is the amber from Chiapas, Mexico, and the Dominican Republic. This may be due to its proximity to the United States and the availability of rare fossiliferous pieces, but a major factor certainly is the exquisite preservation of inclusions, probably the best of any amber. In the Dominican Republic and Mexico, amber was well known to the native peoples. Christopher Columbus apparently received gifts of amber from the Taino people when he landed on the northern shore of the Dominican Republic. In Mexico, amber was carved and used for incense by the Maya, and some indigenous use of it still exists. Mexican amber has been known to North American and European scientists since about 1890, whereas Dominican amber was not scientifically known until about the mid-1940s.

An unusually large piece of Mexican amber, with one surface unpolished and showing deep natural grooves. Length 6.5". American Museum of Natural History (Entomology)

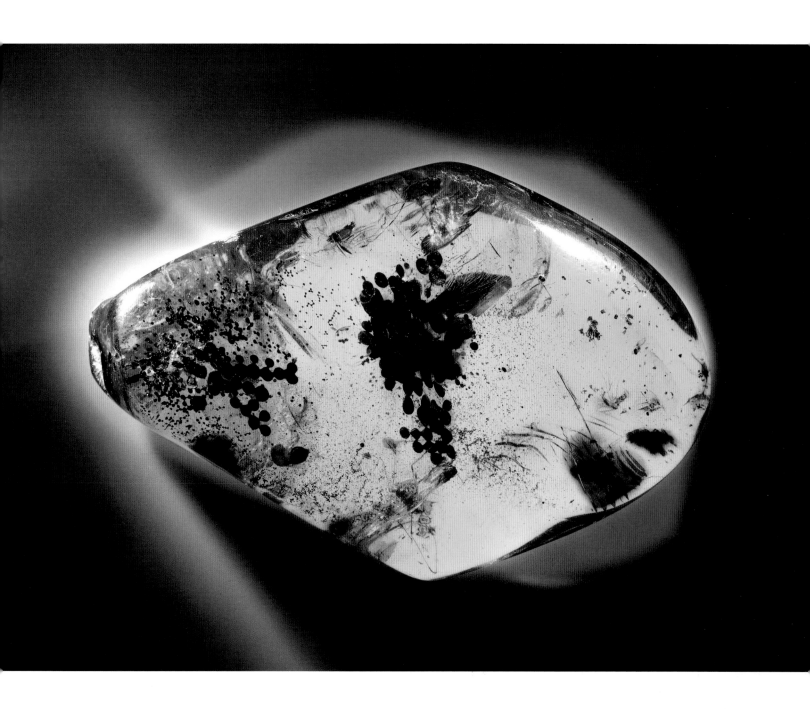

Clear yellow Mexican amber containing
dark bubbles and two small amblypygids,
a kind of arachnid. Length of amber 3.1".
American Museum of Natural History
(Entomology)

Mexican and Dominican ambers were both formed from extinct species of *Hymenaea* trees, although the only one yet described (definitively named) is *Hymenaea protera* from the Dominican Republic. Well before it was named, scientists had discovered that this Dominican amber tree is more closely related to the one African member of the living species in this genus (*H. verrucosa,* a source of African copal) than to any New World species. The Mexican amber tree is apparently most closely related to the living *H. courbaril,* which is widespread in southern Mexico and the Caribbean, all the way down through South America. Identifications of the source of the ambers are based on chemistry and on the whole and partial flowers and leaves in the amber. Dominican and Mexican deposits are approximately contemporaneous, having been formed from around the mid-Oligocene (about 30 million years ago) to the early

Hymenaea courbaril *leaves, flowers, and seedpod. This is a living relative of the extinct trees that gave rise to Mexican and Dominican ambers.*

Opposite: Hymenaea courbaril *tree, Saint John, U.S. Virgin Islands*

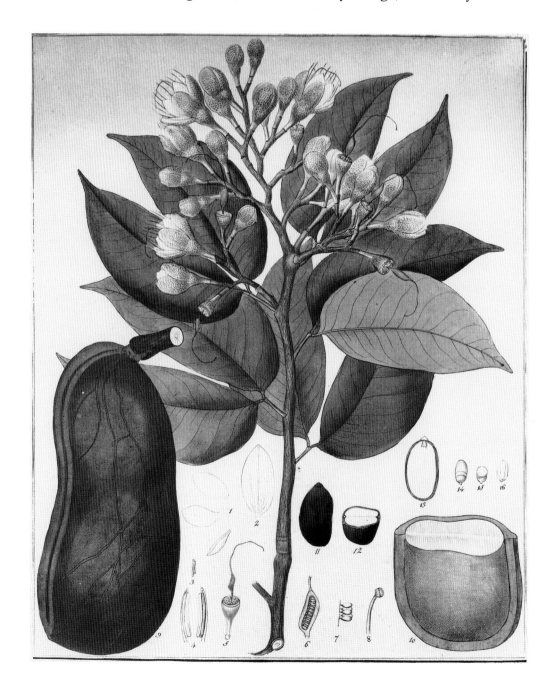

Detail of the small crab in the amber piece at right. Width of crab .2"

Right: Section of amber from Chiapas, Mexico, with a small crab, possibly of the family Grapsidae, in it. An exceptionally rare fossil, this is the only known crab preserved in amber. Private collection

Opposite, above: Large centipede in Mexican amber. Length of amber 2". American Museum of Natural History (Entomology)

Opposite, below: Male ant in a piece of Dominican amber. Length of amber .6". The ant has a metallic shine and is deep red because the body cavity is pyritized. Some amber pieces that have inclusions exposed to the surface are affected this way by dissolved minerals in the surrounding matrix. American Museum of Natural History (Entomology)

Miocene (about 20 million years ago). Very little basis exists for some claims that Dominican amber is 40 million years old.

Fourteen species of *Hymenaea* are found today throughout the Caribbean, tropical South America, and, curiously, the western half of Central America (separated from the eastern half by a central spine of mountains). In the Dominican Republic, *Hymenaea* trees are called *algorrobo,* and the resin is *peruvia* (in Costa Rica, *Hymenaea* is *guapinol,* or stinking toe). Leaves and the large, hard seedpods are studded with tiny pockets of resin, which chemically defend the tree from caterpillars, weevils, and other herbivorous insects. *Hymenaea* produces prodigious quantities of resin from its trunks and branches, sometimes accumulating in large "stalactites." Very large pieces of Dominican amber are sometimes found: one piece of approximately 17.5 pounds is in a shop in Santo Domingo; another, of 15.8 pounds, is in Hamburg, Germany.

Both Mexican and Dominican amber occur in similar settings and are mined in similar ways by locals. Generally the amber is found because a landslide along a steep slope in the mountains exposes veins of black lignite. If the lignite contains amber, it is gradually extracted by digging along the vein with picks and shovels. In a rich seam, several pounds of amber can be extracted in a day. When the veins extend deep into the mountain, the diggings evolve into tunnels, sometimes 100 to 200 feet long. This is especially true in Mexico; in the Dominican Republic, tunnels are dug only in the La Toca group of mines (most other diggings are broad, deep pits). Water accumulates in the tunnels and must be baled or pumped out. Even so, the tunnels sometimes collapse, as has happened in the Dominican La Toca mines.

After digging, the miner takes the material into bright sunlight, washes it, and chips a small piece off one end to expose a clear window. If the window reveals any special, large organism inside, it is reserved for special bargaining with amber dealers. In the Dominican Republic, the dealers centered in Santiago and Santo Domingo have corps of polishers, some of them children, who remove the rind from the crude amber and polish it, generally following the natural contours. Thousands of pieces are processed each week in the larger amber shops, all sorted according to size and whether they have rare inclusions or not. Small, barren pieces are used for necklaces, bracelets, and earrings. In Mexico, the grinding and polishing is more of a cottage industry, but even here the choicest, rarest fossil pieces make their way to an international market of amateur collectors and museums.

Scientific study of the organisms has revealed, particularly for Dominican amber, an exceedingly rich extinct biota, our knowledge of which is based on collections of Dominican fossils in the Museum für Naturkunde in Stuttgart, the Smithsonian Institution, and the American Museum of Natural History. The paleontology department at the University of California, Berkeley, has an intensively studied collection of Mexican amber fossils. There also exist several superb private collections of Dominican and Mexican amber fossils.

Amber mining in the Dominican Republic

Opposite, clockwise from left: Miner in an open pit at the Los Cacaos mines, Dominican Republic

The famous La Toca group of mines, near the ridge of a steep slope. Ramón "Rubio" Martinez, a famous amber dealer, is in the foreground.

Open pits and huts at the Los Cacaos mines. These mines are acclaimed for the blue amber they yield.

Three men working in a deep pit at the Los Cacaos mines

Overleaf, left: Chunk of Dominican amber in its siltstone matrix. Length of amber 2.1". American Museum of Natural History (Earth and Planetary Sciences)

Overleaf, right: An unusual piece of Dominican amber with the opaque, milky swirls more commonly seen in Baltic amber. Length 4.6". American Museum of Natural History (Entomology)

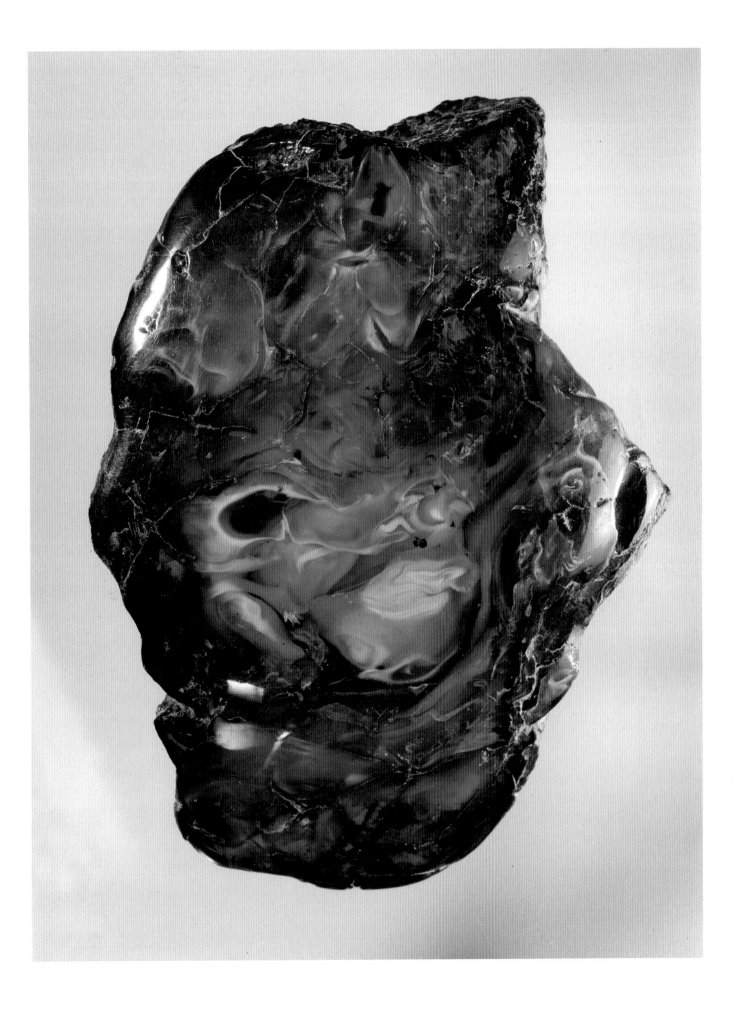

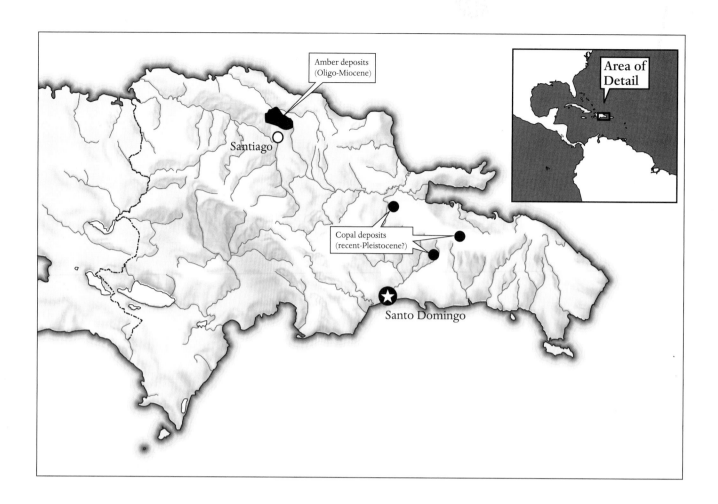

Area of Detail

Amber deposits (Oligo-Miocene)

Santiago

Copal deposits (recent-Pleistocene?)

Santo Domingo

Amber mining in the Dominican Republic

Opposite: Two pieces of Dominican amber with unusual inclusions. The long piece contains wood (length 2.3"); the smaller one has opaque milky clouds. American Museum of Natural History (Entomology)

Unlike the Baltic amber, Mexican and Dominican amber rarely occurs in milky, opaque forms; it is usually very transparent. Insect and other inclusions in Dominican and Mexican amber also rarely are obscured by a milky substance, although organisms in Mexican amber are frequently distorted by compression.

Amber from Mexico is concentrated around the town of Simojovel, in the southern state of Chiapas. Dominican amber comes from one of about thirteen groups of mines approximately 50 miles northwest of Santiago, arising from the La Toca Formation at least 1,500 feet high in the Cordilleira Septentional, and two other mines considerably north of this group, not far from Puerto Plata. Some mines are renowned for their distinctive colors, although color is hardly consistent. Some amber from the Los Cacaos mine is the bluest of probably any known. Amber from Palo Alto is famed for its clear yellow hue and that from the La Toca mines for its deep red color, although both colors occur in both mines. Occasionally, miners find pieces of smoky, greenish amber. Such colors are found in Mexican amber as well.

A light, almost clear fossil resin (copal) in the Dominican Republic, often sold as amber, comes from the eastern towns of Bayaguana, Cotui, Comatillo, and Sierra de Agua. As in all copals, it becomes heavily crazed in several years. One scientific study estimated this copal to be 15 million years old, but carbon dating has revealed that at least some of it is only several hundred years old.

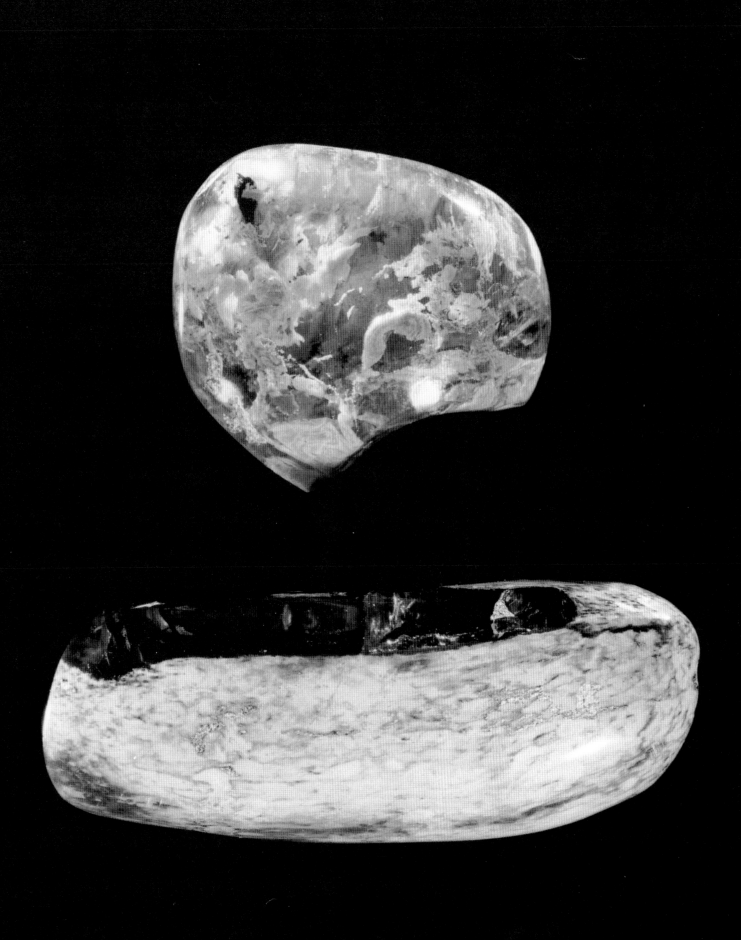

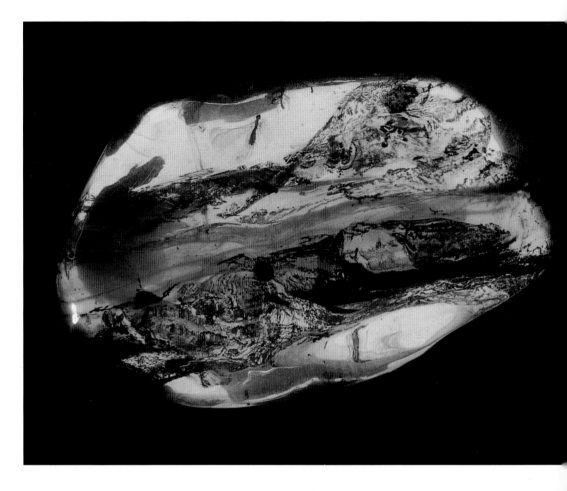

This page: A piece of Dominican amber lighted completely from behind (above), and with long-wave ultraviolet and some reflected white light (below). Under ultraviolet light, the fluorescing amber appears very dense, and visible are many more flow lines than are seen in transmitted light. The piece contains some termites. Length of amber 2.9". American Museum of Natural History (Entomology)

Opposite, above: A leaflike pattern of pyrite, or "fool's gold," lying on a fractured surface in Dominican amber. Length of amber 1.3". Private collection

Opposite, below: A piece of deep red Dominican amber, part of it highly polished, the remainder with a natural surface of deep fissures. Length 3.8". Private collection

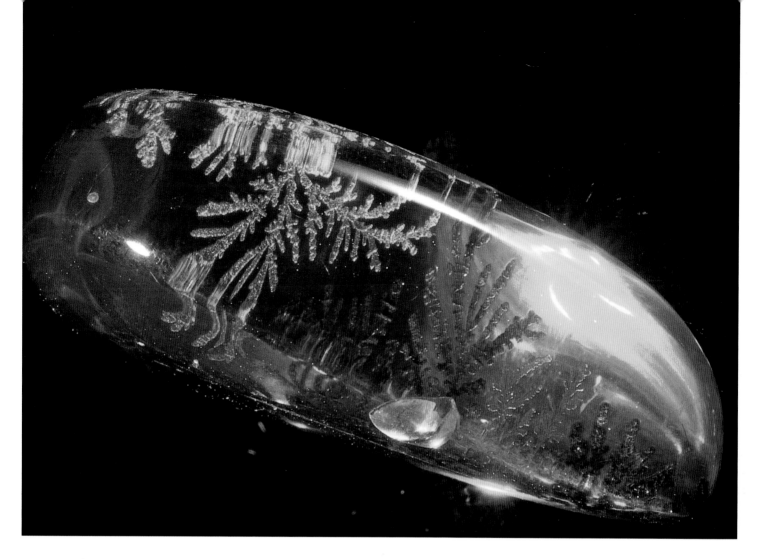

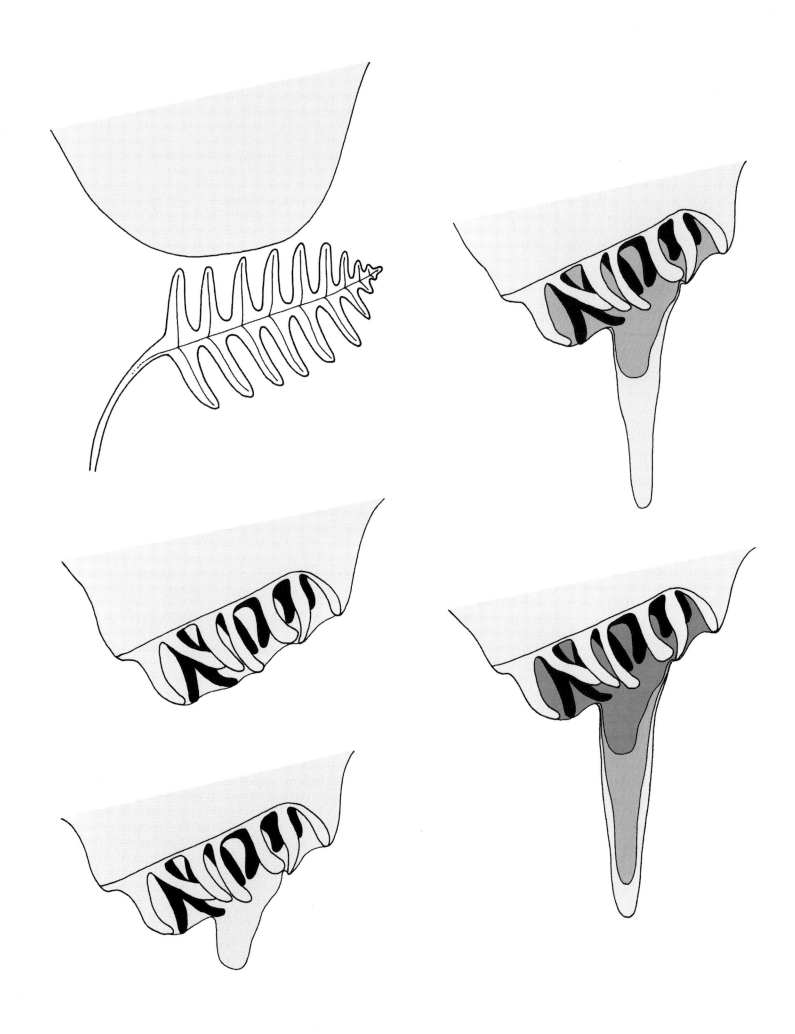

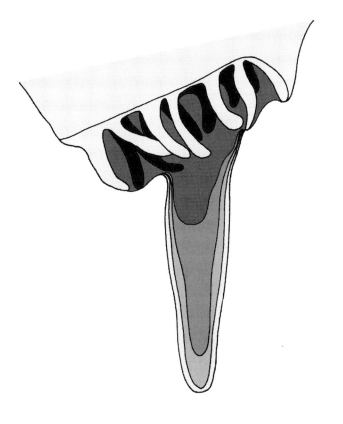

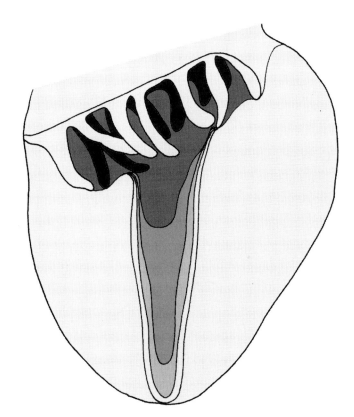

*Opposite and left (running in columns):
Reconstruction sketches showing how the
piece below was formed*

*Fern in 25-to-30 million-year-old Dominican
amber. The fern is curled up, and "stalactites"
of amber are hanging from it. Length of
amber 2.2". Private collection*

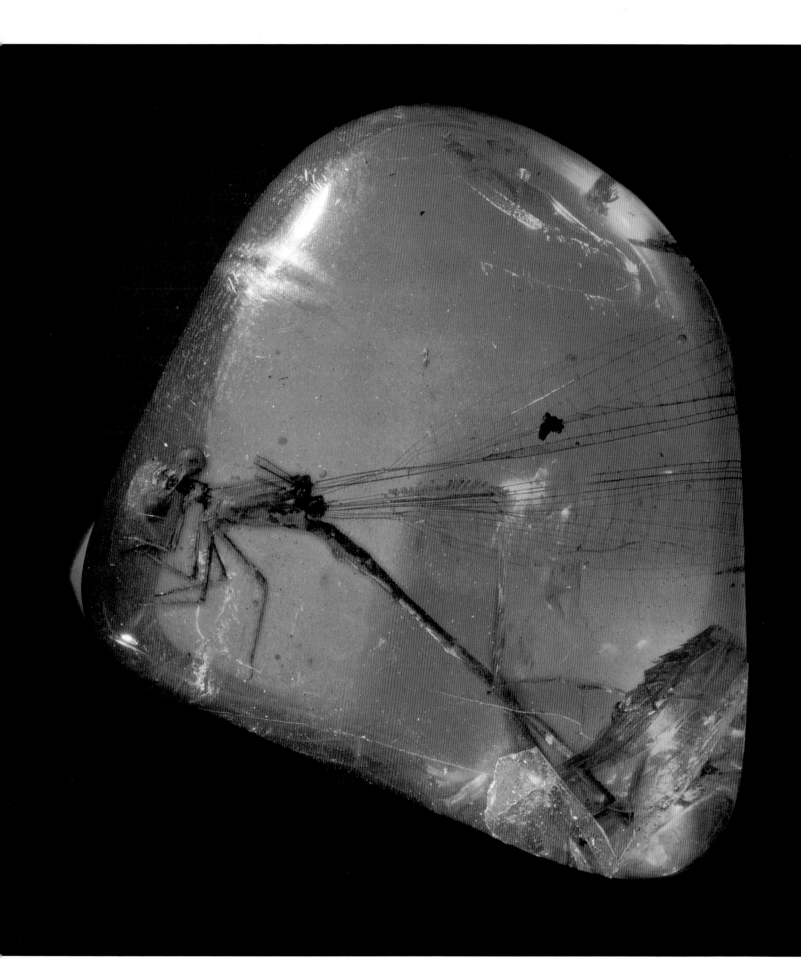

FROZEN IN THE ACT

An ant beneath a poplar found,
An amber tear has covered round;
so she that was in life despised,
in death preserved, is highly prized.

In the bright tear Phaëthon's sister shed
a bee is seen, as in its nectar, dead.
Its many toils have earned a guerdon high,
in such a tomb a bee might wish to die.

—Martial, *Epigrams* (Books vi.xv; iv.xxxii)

*I*n the exceptional circumstance, a fossil is found that reveals something of
its life history and habits. The fabulous ichthyosaurs from Württemberg,
Germany, are preserved in a slate so fine that one can see that they gave live
birth to their young and, sometimes, can discern what they ate. Fish from the
Santana limestone of Brazil (about 110 million years old) are preserved in
remarkable three-dimensional concretions, with the muscle bundles entirely
replaced by minerals with their shape still intact. Occasionally one is found
replete with the little bodies of the shrimp that it dined upon. For organisms
as delicate as insects, "freezing" a prehistoric moment requires exceptional
preservation, which amber provides. Ambers have preserved the various
developmental stages of some insects, prey and plant hosts, parasites,
commensals, as well as exhibitions of defensive and social behavior. Most
of the examples will be taken from the 25-to-30-million-year-old Dominican
amber because the diversity of its inclusions allows the most complete
reconstruction of the ancient forest.

Dispersal of a species is necessary if opposite sexes are to meet and
reproduce. In situations where the ability to get around is limited, as with
an arthropod that lacks wings, novel solutions are required. One of the most
interesting is *phoresy,* or hitching a ride on another animal. Phoresy is common
in mites that disperse from, say, mushroom to mushroom on a fly. Phoretic
mites on various flies are preserved in amber, but probably the best examples
are some sweat bees (halictids) with dozens of tiny mites still latched on for
one last, fateful ride.

*Fossils in this section are preserved
in 25-to-30-million-year-old amber
from the Dominican Republic, unless
otherwise noted.*

*Opposite: A damselfly, slender relative
to dragonflies. Length of amber 1.8".
Private collection*

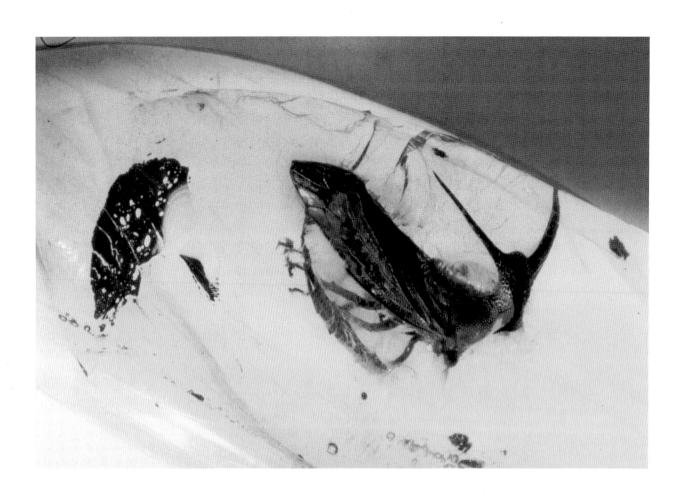

Opposite, above: Membracid treehopper. The part of its body just behind the head is prolonged into a spine with three thorns. Length of amber 1.4″. American Museum of Natural History (Entomology)

Opposite, below: Leaf beetle (family Chrysomelidae) that exuded a stream of noxious bubbles in an attempt to defend itself from the resin. American Museum of Natural History (Entomology)

Above: Stick insect, or phasmid. Length of amber 3.4″. Private collection

The most curious example of phoresy in amber involves pseudoscorpions, tiny renditions of scorpions without the stinger. Many pseudoscorpions live under bark or among cracks in bark, where they feed on mites and other tiny arthropods. Sometimes one is found with a claw clamped onto a braconid wasp or tipulid flies, but in Dominican amber they are attached most often to wood-boring platypodid (ambrosia) beetles. We know today that some pseudoscorpions live in the galleries made by wood-boring beetles. When the pseudoscorpions disperse, they latch onto the first beetle that comes along, which may take them to another tree like the original.

Right: Wood-boring platypodid beetles, with the sawdust plugs that they pushed out of their tunnels in wood. The beetles were probably attacking a Dominican amber tree. American Museum of Natural History (Entomology)

Below: Wood-boring platypodid beetle, with a pseudoscorpion latched onto it with one of its claws. American Museum of Natural History (Entomology)

Opposite: Pseudoscorpion. Length of amber .4". American Museum of Natural History (Entomology)

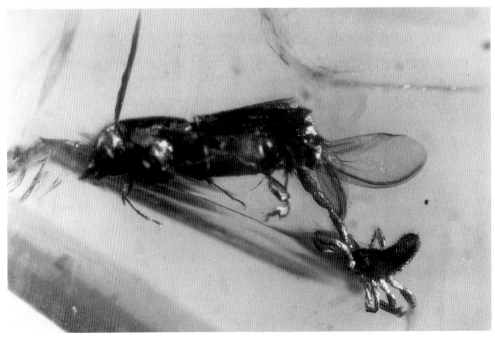

Insects have a plethora of strategies for reproduction; some are even captured in amber. Among the short-lived (usually aquatic) insects, such as many gnats, mosquitoes, midges, and mayflies, the males aggregate into sometimes huge swarms. Thousands, or millions, of males fly about in one spot, into which females fly to become mated, thus insuring that the sexes rendezvous during their brief life span. Copulation may take hours, as in the case of a bibionid midge, *Plecia nearctica,* the "love bug." Male swarms, or portions of them, in Dominican amber are most commonly of scatopsid midges, but examples of empidid flies, dolichopodid flies, mycetophilid midges, tipulid flies, and termites also appear. Termite swarms caught in amber usually have a jumble of wings among the bodies, since termites easily shed their wings after landing. Occasionally a mating pair of midges is caught in the resin. One piece of Baltic amber even contains a pair of mating spiders. The oldest mating pair, of any kind of animal, are sciarid midges in 125-million-year-old Lebanese amber.

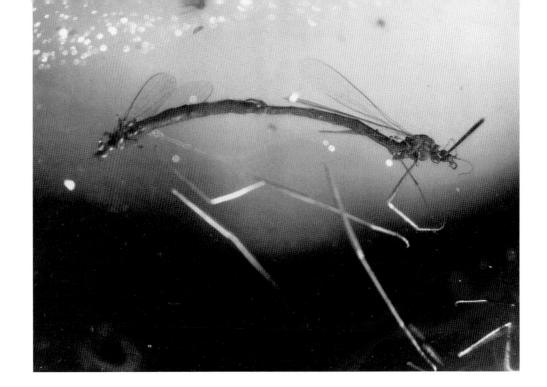

From top to bottom: Mating crane flies. Length of amber 1.2". Private collection

Detail of coupled pair above

Midge trailing a string of her eggs. American Museum of Natural History (Entomology)

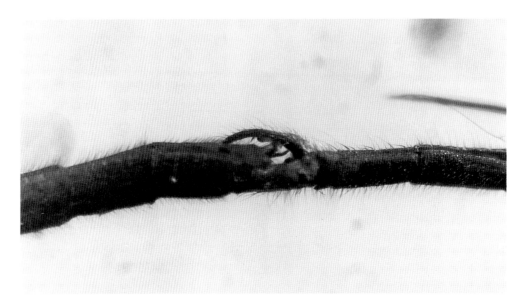

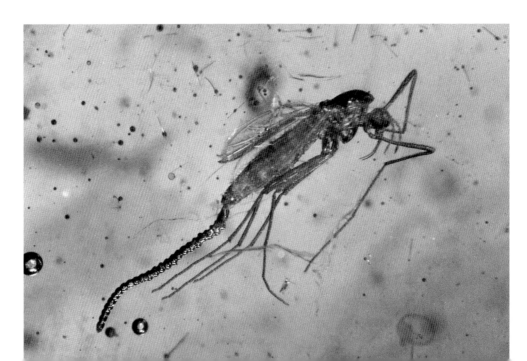

Many flies reflexively lay eggs when they die, which explains why some female flies in amber have eggs just behind them. This is seen in Dominican amber most commonly in the little drosophilid fruit flies but also with some midges, such as the chironomid trailing a string of her eggs. Occasionally a decaying insect or other unidentified decaying tissue has clumps of fly eggs on it, which never quite hatched before resin engulfed them. One exquisite aspect of insect eggs in amber is that the intricate geometric sculpturing of the eggshell is still apparent. In one case, a young larva was caught emerging from its egg. In several other cases, loosely woven silken cocoons of spiders still have the embryos or newly hatched spiderlings within. Various kinds of larvae occur in amber, but the choice specimens are of a larva in some interesting situation, such as with silken cases. Bagworm, or psychid moth, caterpillars carry around them a bag of silk, to which has been sewn many bits of leaves and twigs of the plant they were feeding upon. It camouflages them extremely well, but it did not

Wood gnat emerging from its pupal case. American Museum of Natural History (Entomology)

Overleaf, left: Metalmark butterfly (family Riodinidae). Length of amber 2". Private collection

Overleaf, right: Large inchworm moth (family Geometridae). Length of amber 2.2". American Museum of Natural History (Entomology)

prevent a few from being engulfed in resin millions of years ago. Tineid moth caterpillars do a similar thing on a smaller scale, and generally they use their own frass (insect feces) for concealment. The ones preserved in amber were probably grazing upon the woody polypore (bracket) fungi that grew on the Dominican amber tree.

Social insects have colonies of hundreds to hundreds of thousands of individuals, divided into castes for defense, egg-laying, and working. Sometimes the workers are subdivided into nurses and major and minor workers. One just needs to see the huge termite mounds on an African savanna, or an army ant swarm in a South American jungle, to appreciate how social insects are among the most ecologically important groups of animals, and generally the most conspicuous insects. In Dominican amber the social insects are ants (the most common of all the inclusions), termites, stingless bees, and, rarely, paper wasps. Various castes are preserved, including the bizarre workers of *Zacryptocerus*, which plug the entrance of the colony with their flat heads. To enter, an ant must tap properly with its antennae. Another weird caste is the nasute soldier of some termites. Nasutes have heads shaped like a bottle, from which they spray a sticky substance at intruders. Of course, portions of colonies are

Opposite: Paper wasp. Length of amber .8". American Museum of Natural History (Entomology)

Ants caught while attempting to carry their larval brood to safety. Museum für Naturkunde, Stuttgart

Portion of an Azteca ant colony. This piece contains about two hundred ants. Length of amber 1.5″. American Museum of Natural History (Entomology)

occasionally engulfed by resin, sometimes with hundreds of ants (one piece in the Stuttgart museum contains about 2,000 ants). Very rarely a piece is found in which the workers were caught trying to carry the brood to safety. The only portions of the actual nest are several cells from the paper wasps. However, the middens and remains of termite and ant nests with their original owners abound in amber.

The middens of the ant colonies provide excellent clues as to what the colony was feeding upon, such as assorted body parts of other insects. Every social insect colony today has "guests," some wanted, some not. Some are parasites; others—the *inquilines*—merely live off the scraps in the nest. In Dominican amber, there are nicoletiid silverfish, certain tiny staphylinid, limulodid, and paussine beetles, all superbly adapted for living undetected among (or at least tolerated by) ants in their colony. And there are the parasites of the ants, too, such as the bizarre twisted-winged parasites, and various scuttleflies of the Phoridae family. The social insect colony, then as now, is a cosmos of ecological relationships.

Many predatory beetles, lacewings, robber flies, sucking bugs, spiders, and even mantises and damselflies have been trapped in amber, but a predator caught in the act with its prey is rare. Sometimes a piece contains spiderwebs with the victim (generally a tiny, frail gnat) snagged on a thread. Never has a piece been found with the spider still resident. One piece in the American Museum of Natural History has a jumping (salticid) spider grappling with its millipede prey. A famous piece of Baltic amber has a *Ptilocerus* assassin (reduviid) bug in it, along with the husks of the ant prey that the bug sucked dry. Living relatives of this bug today lure ants from their nests with the scent from a special gland, then they

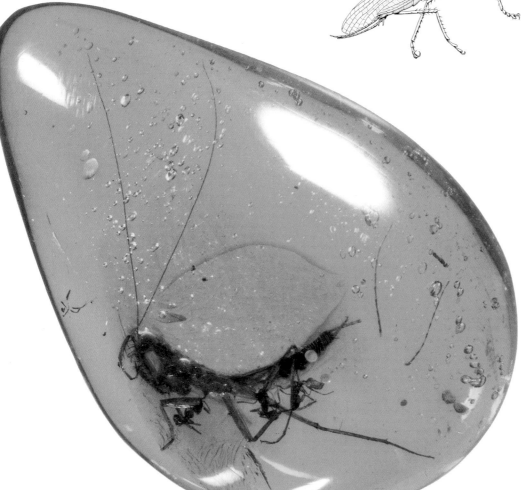

Top: A mantis look-alike: mantis-pid lacewing in Dominican amber. Length of specimen .9". Private collection

Above: A rare adult praying mantis in Dominican amber. Private collection

Left: A praying mantis, attacked by ants, carried them to its resinous tomb. Length of amber 1.2". Private collection

Opposite: Dominican amber with two amblypygids (whip scorpions) and various small insects in it. Width of amber 3". Private collection

Detail of amblypygid in the piece opposite (and on the jacket front), showing insect prey still caught in its spiny, basketlike jaws

Jumping spider embracing its millipede prey. American Museum of Natural History (Entomology)

Above: Spider. Length of amber 1.6″. Private collection

Parasitic fly, Stylogaster, with a rapier egg-laying appendage. Living species of this genus today parasitize cockroaches. Private collection

impale the ant with their sharp beaks. A similar bug is found in Dominican amber, but it probably fed on the stingless bees, *Proplebeia dominicana,* that were very common in the Dominican amber forest. The foreleg of each bug has a large droplet, which it must have used as a sticky snag for the bee.

The parasitic insects are, likewise, very common in Dominican amber, especially the various minute wasps. (Entomologists refer to insects that live on and eventually kill another, host insect as a *parasitoid,* to distinguish them from true parasites, which do not kill their hosts.) Few direct evidences of parasites and parasitoids exist in amber. Mites are the most common type of parasite found on insects. Water mites are found on the adults of some aquatic insects, such as caddis flies and chironomid midges. Some tiny moths harbor erythraeid mites, and small drosophilid flies occasionally have large macrochelid mites (proportional in size to a human with a watermelon attached). The most gruesome are the nematode worm parasites; in one piece of Dominican amber from the Stuttgart museum, a huge mermithid nematode can be seen emerging from its midge host; the nematode must have taken up most of the host's body. Several pieces of Dominican amber have leafhopper nymphs with a large black sac attached near the abdomen. This is a dryinid wasp larva (adult dryinid wasps also are in the amber). A personal favorite is a Dominican amber specimen containing tangled strands of spider webbing. Dangling along the strands is a row of seven tiny cocoons; the wasp larvae that spun these cocoons parasitized the spider on whose web the cocoons are now preserved. Adult wasps emerged from all but three of the cocoons.

Above: Tiny twisted-winged parasite (order Strepsipteran) of ants. Only .5 mm long, it is remarkably similar to a present-day species. American Museum of Natural History (Entomology)

String of tiny cocoons suspended from a spider web. The cocoons are from wasp larvae that parasitized the spider that spun the web. American Museum of Natural History (Entomology)

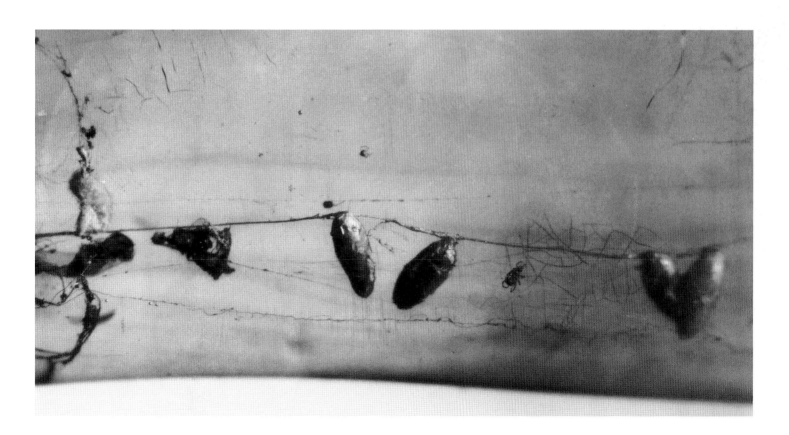

Ancient Communities:
Reconstructing the Ancient Dominican Amber Forest

Pretty! in amber to observe the forms
Of hairs, or straws, or dirt, or grubs, or worms!
The things, we know, are neither rich nor rare,
But wonder how the devil they got there.

— Alexander Pope, Epistle to Dr. Arbuthnot

Small menagerie of 217 insects, spiders, and plants. The "spray" of tiny insects is collembola, or springtails. The flowers and stems are from an acacia. Length of amber 1.5". Private collection

Preservation in amber is as biased as any other kind of fossilization, particularly when it comes to size. Large insects generally have the strength to free themselves from sticky sap, so it is very rare to find large beetles, dragonflies, grasshoppers, and the like in amber. In fact, the longest insects discovered in amber are damselflies (about two-and-a-half inches long) that could not free themselves from the resin because they are so delicate and thin. Likewise for plants: only those flowers and leaves small enough to be blown about in the wind and encapsulated by a stream of resin are the ones that are preserved. This lilliputian bias in fossilization is made up for, though, by the lifelike detail preserved in the amber

and the sheer diversity of tiny organisms. It is well documented that, at least for insects, the number of species increases tenfold for a tenfold decrease in body size. And the more species that are preserved, the more complete is a reconstruction of the amber forest. A fascinating insight is revealed by discovery of amber pieces with a menagerie preserved inside, forming a true snapshot of a tiny part of the community. Some of these pieces display striking diversity. One in the Stuttgart museum, for example, has in it some two hundred individual arthropods belonging to twenty-two families.

The most direct signs of the Dominican amber forest are the assortment of flowers, stems, leaves, seeds, and even tendrils preserved in the amber. Living on the trunks of the Dominican amber tree, as happens now, were mosses, liverworts, and the occasional mushroom. Living among the amber trees were mimosoid trees like acacias. Flowers of the families Bombacaceae (balsa and baobab family), Euphorbiaceae (euphorbs, such as cassava and poinsettia), Hippocrataceae, Leguminosae (pea family), Meliaceae (mahogany family), Myristicaceae (nutmeg family), and Thymeliaceae have all been identified in the amber.

Inference of the ancient amber forest can be made based on the myriad insects preserved in amber. Adult and immature insects occupy various niches in freshwater (and, occasionally, in brackish and salt water), soil, and decaying wood; as parasites of other insects and of vertebrates; and feeding on the entire

Map of the piece opposite

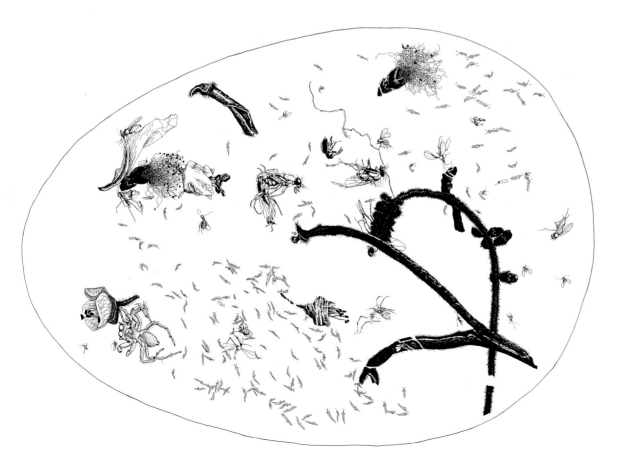

A menagerie piece, only 1.1" in diameter, containing sixty-two whole or partial insects representing five orders and fourteen families. Some of the inclusions are covered with mold. American Museum of Natural History (Entomology)

Opposite: Reconstruction of the ancient Dominican amber forest. The numerous life-forms preserved in this amber allow a detailed re-creation of what the forest probably looked like, including the inhabitants of the forest floor, living under bark and in the amber trees and on the plants growing near the amber trees. Everything in the reconstruction is either supported by actual fossils in Dominican amber or inferred on the basis of host plants for plant-feeding forms of insects.

array of fungi, flowers, and leaves. Some insects are dedicated to a particular kind of plant; for example, monarch butterfly caterpillars concentrate on milkweeds, others are great generalists. If an insect in amber has living relatives feeding exclusively on a particular genus of tree, we can be fairly certain that the extinct species fed on an extinct species of the tree. For example, we are fairly certain that fig trees lived in the amber forest, even though we have no direct evidence. These trees are renowned, among other things, for the "flying buttresses" that help support their gargantuan proportions, in contrast with the minuscule insects that pollinate them. Living in the figs (which are actually an unusual kind of inflorescence called a *synconium*) are agaonid wasps about a millimeter long. Each species of fig harbors a specific species of wasp, and the wasps are found nowhere else. Dominican amber has fossilized several of these fig wasps.

Although we have no direct fossil record from the Caribbean, we also know that palms were in the Dominican forest, based on the thausmastocorid palm bugs and certain kinds of weevils in the amber. Ultimately, a comprehensive study of the array of plant bugs, plant hoppers, leafhoppers, whiteflies, scale insects, leaf beetles, and moths in Dominican amber will reveal what this 25-million-year-old forest was like. As of now, we know that open areas occurred in at least some parts of the forest, not only because of a few grass spikelets found in Dominican amber, but also since there are lygaeid bugs in it as well. Bromeliads

Opposite: Hymenaea *leaf, from the amber tree. Length of amber 2.1". American Museum of Natural History (Entomology)*

Above: Hymenaea *flower, from the tree that formed the amber. Length of amber 1.6". Private collection*

Below: Winged seed. Private collection

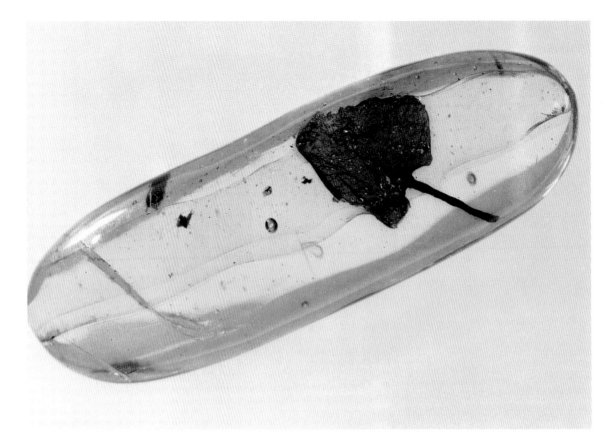

Petal from a Hymenaea *flower, in Mexican amber. Length of amber 1.3". American Museum of Natural History*

Small flower with a thorny stem, its pollen spreading into the once-liquid resin. Length of inclusion .9". Private collection

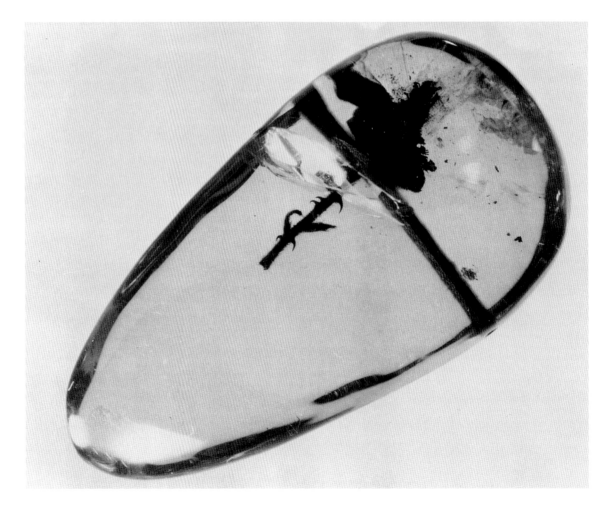

were nestled among the branches of the Dominican amber trees themselves. A species of butterfly in Dominican amber (a metalmark) and its caterpillars probably fed upon the bromeliads. Living in the little ponds that accumulate in the center of the bromeliads, no doubt, were mosquitoes, predacious diving beetles, and perhaps even the small frogs found preserved in the amber.

The wood of the Dominican amber tree was infested with various insects. The most common sign of this is frass, or the tiny pellets of insect feces. Frass is in all amber that contains insects, and, at least for Dominican amber, it is not at all uncommon to find dozens of frass pellets, which probably rained down into the resin from an opening in an adjacent insect nest, in a single piece. Most frass in Dominican amber appears to have come from termites. We can even surmise that the Dominican amber forest landscape was dotted with large carton nests of *Nasutitermes* termites. Colonies of these termites today build intricate oblong nests—attached to tree trunks or hanging from branches—that have a composition like brittle papier-mâché. As do most termites, the nasute worker termites construct thin galleries, in this case meandering all over the tree to the ground, through which they and the soldiers march unexposed.

Colonies and nests that were less conspicuous, but which probably had a much greater ecological impact than those of any other social insects, were those of the giant *Mastotermes* termites. The extinct species from Dominican amber, *M. electrodominicus,* had winged reproductives nearly an inch-and-a-half long. A similar extinct species, *M. electromexicus,* exists in Mexican amber. The only living species in this genus is in Australia. The Australian species constructs large subterranean colonies, generally at the base of the trees whose wood they are consuming, and they are voracious.

Another enemy of the Dominican amber tree was a plethora of wood-boring and bark beetles, also called ambrosia beetles. Numerous species in the families Platypodidae and Scolytidae occur in Dominican amber. The beetles today excavate tunnels and galleries throughout heartwood, or into the surface of the heartwood just under the bark (bark beetles). Sawdust produced from the tunnel excavations is pushed out through the tunnel entrance and compressed into cigar-shaped plugs. Such plugs are preserved in amber along with the beetles. The beetles attack a living or injured tree but do not actually kill it by their boring. A fungus specific to each species of beetle is inoculated into the wood, where it grows to carpet the galleries. The beetles feed on this fungus (their "ambrosia"), and it is the fungus that kills the tree. We know that some trees today secrete excessive amounts of resin to trap beetles invading their wood; the beetles in amber are evidence that the strategy works, at least somewhat. Similarly, it has been thought that the Dominican amber was produced in such large amounts to offset massive outbreaks of bark and wood-boring beetles.

Judging from the incredible variety of little brown beetles, fungus gnats, certain kinds of rarely collected acalyptrate flies, and various other kinds of termites, all associated today with rotting wood and the fungi that decompose

Stem and leaflets of an acacia plant.
Length of amber 2.2". American
Museum of Natural History
(Entomology)

Opposite, above: Liverwort. Length
of amber 2.3". Private collection

Opposite, below: Small mushroom.
Length of amber 1.5". Private
collection

it, the ancient forest must have been littered with dead and dying trees, stumps, and logs. The full array of life forms dwelling under the loose bark of dead and dying amber trees included mites, predatory beetles, aradid bugs, earwigs, bristletails and silverfish, centipedes, pseudoscorpions and true scorpions, and fearsome-looking amblypygids. Resin streaming out of the trees must have accumulated in such masses that globs dropped to the ground, engulfing organisms living among the leaves and soil, such as springtails (Collembola), pill bugs (isopods), various millipedes, burrowing bugs (Cydnidae), and an assortment of tiny snails. Even earthworms and a mole cricket have been discovered preserved in the amber.

We find evidence of freshwater adjacent to the forest: perhaps streams flowed through, as indicated by the damselflies (Anisoptera) in amber, as well as several species of mayflies (Ephemerida), a stone fly (Plecoptera), pond skaters (Gerridae), and an assortment of adult caddis flies (Trichoptera) and midges whose larvae are aquatic. Of the dozens of beetle families in Dominican amber, four are aquatic, and for one of these (the Helodidae), even the aquatic larva is preserved.

The most highly prized specimens in Dominican amber are the small anole and sphaerodactyl gecko lizards and small *Eleuthrodactylus* frogs. Today in the Caribbean are hundreds of species of these kinds of lizards and frogs, more than all other kinds of land vertebrates. Traces of vertebrates such as feathers, hair, and sloughed reptilian skin also occur in the amber. One of the feathers has been identified on the basis of its microscopic structure as definitely from a woodpecker,

X-ray positive of the gecko lizard in the piece at right and on page 110, showing tiny bones in the legs and feet. The vertebrae are jumbled; the lizard broke its back in several places, perhaps while struggling to free itself from the resin. The leaf is not detected by X rays.

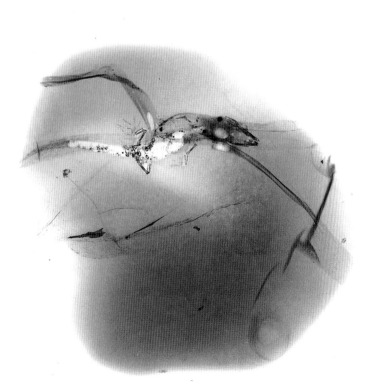

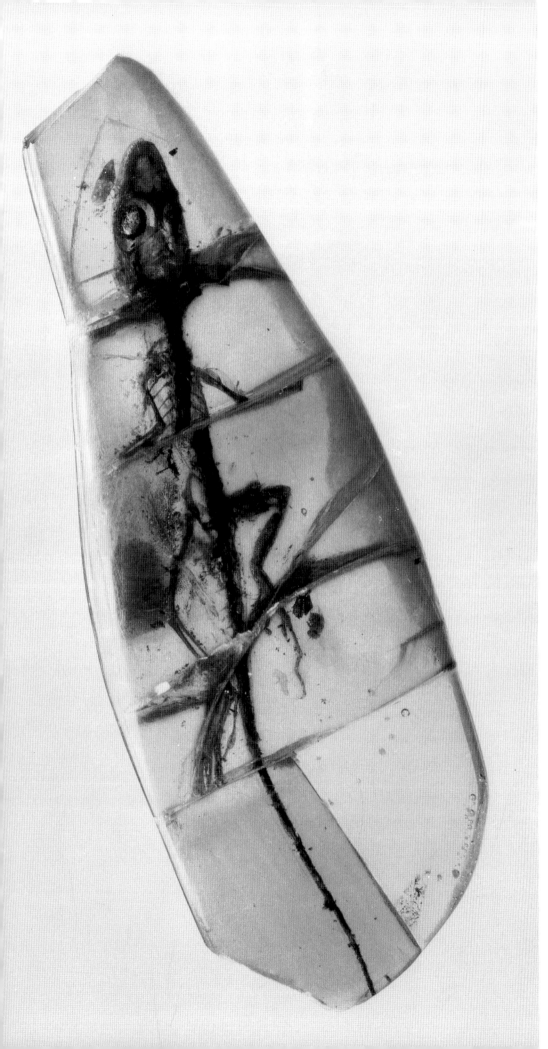

Small anolis lizard, partially skeletonized. The piece was cracked and glued back together by workers in the Dominican Republic. The light section is a substitute piece. Length of amber 2.2". American Museum of Natural History (Entomology)

*Opposite: Small gecko lizard,
Sphaerodactylus, near a leaf section
that has been chewed, probably by a
leaf-cutter bee. Length of amber 1.7".
Private collection*

*A complete frog and, above it, a
decaying frog, Eleuthrodactylus, part
of whose backbone is easily visible.
Surrounding the decaying frog are
dozens of fly larvae. How such a piece
was formed is perplexing, although
one thought is that a predator dropped
its frog prey into resin. Length of
amber 2.3". Private collection*

*Reptilian skin shed by a snake or large
lizard. Length of amber .8". Private
collection*

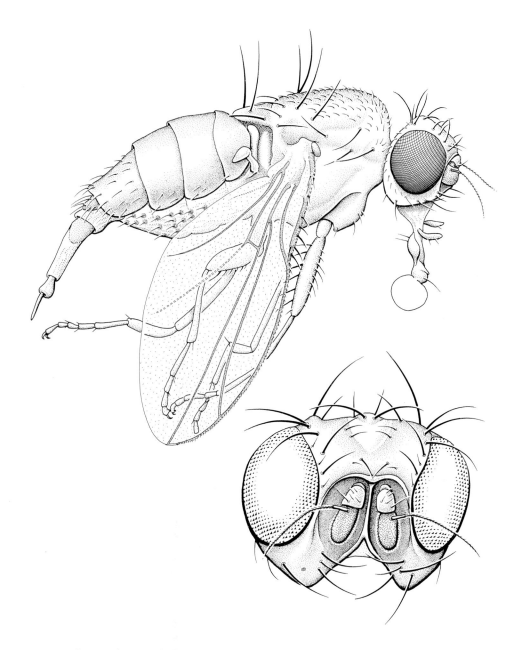

quite similar to the Antillean piculet. Some hair has been identified as perhaps from a sloth, although few diagnostic features exist to support that conclusion. Other than a fragment of a ground sloth skull from Cuba, and a few other fragments from Puerto Rico, no other Oligo-Miocene land vertebrates from the islands are known. All the others are fossils of monkeys, sloths, rodents, and the like from certain periods in the Pleistocene, some 20,000 years old and younger. Yet Dominican amber reveals that vertebrates did roam the amber forests, and it was a fauna not unlike what occurs there today. As with the flora, we have both direct and indirect signs of its presence.

More varieties of bloodsucking arthropods appear in Dominican amber than there are kinds of vertebrates, which attests to a very diverse vertebrate fauna. In a few cases, we can even make some specific inferences as to whose blood the arthropods fed upon. The minute *Forcipomyia* and *Culicoides* no-see-ums fed

Above: Detail of a remarkable piece of amber containing a small swarm of phlebotomine sand flies with mammalian hairs and debris, perhaps from the nest of the mammal. The females are bloated, probably with the blood of the mammal whose hairs are in the amber. Museum für Naturkunde, Stuttgart

Right: Tick. American Museum of Natural History (Entomology)

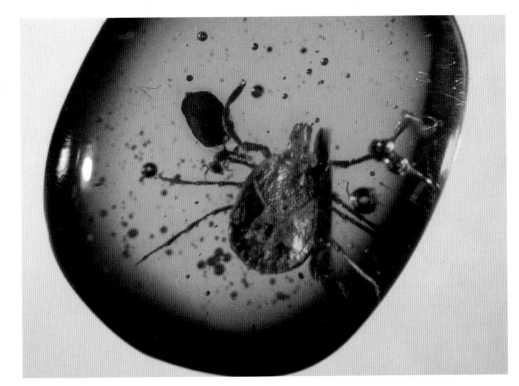

Left: Female mosquito. American Museum of Natural History (Entomology)

Below: An exceptionally rare flea. Museum für Naturkunde, Stuttgart

on mammals and birds, as did the mosquitoes and phlebotomine sand flies. A unique piece in the Stuttgart collection contains numerous fine hairs, debris (perhaps from the mammal's nest), and a small swarm of phlebotomine midges. Most of the phlebotomines are females, and some are bloated. They probably are replete with the blood of the mammal whose hair is preserved in the amber. Some mosquitoes and phlebotomines may have fed on larger reptiles and amphibians. A species of corethrellid biting midge in the amber probably fed on the *Eleuthrodactylus* frogs also preserved in the amber. Similar midges today, in fact, are collected by attracting them to recordings of frog calls. The species of *Stenotabanus* horseflies and *Amblyomma* ticks most likely fed on mammals. For at least one rare specimen, a tick is preserved in a piece with two hairs, indicating that the tick almost certainly fed on a mammal. The few fleas in Dominican amber may have fed on birds. One tiny little fly in the amber, *Meonura,* may have fed on birds, perhaps the same woodpecker species whose feathers were identified, since living species of *Carnus* flies (their close relatives) have a predilection for woodpeckers.

Intricate Preservation

The Spider, Flye, and ant, being tender, dissipable substances, falling into amber, are therein buryed, finding therein both a Death, and Tombe, preserving them better from Corruption than a Royall Monument.

—Francis Bacon

A 1982 study published in the journal *Science* reported on organelles and other subcellular structures from a bit of tissue in a 40-million-year-old fungus gnat in Baltic amber, examined with an electron microscope. It was widely acclaimed as opening the door for future research on the preservation of amber fossils, including ancient DNA. Incredibly, such a result had been reported almost eighty years earlier, using conventional light microscopy and histology. That 1903 report, by Nicolai Kornilowitch, unfortunately was published in a local journal, and in Russian, so it remained almost totally obscure. Kornilowitch reported that dried tissue taken from insects in Baltic amber possessed banding patterns typical of modern muscle tissue. Truly great ideas and results in science are often too far ahead of their time. No one knew of the existence of DNA in Kornilowitch's time, but the 1982 study at least prompted the question, If intracellular membranes and organelles could be preserved, why not chromatin, and even DNA? (As will be discussed later, that study did slowly stimulate interest in extracting ancient DNA from amber fossils, but it was a technological break-through that served as the real driving force.) Since 1982, several comprehensive studies have revealed even more consistent and unexpectedly lifelike preservation of soft internal tissues of insects and plants in amber.

An insect specimen cannot be "exhumed" from amber by melting away the surrounding amber, since amber will melt only under temperatures and pressures so high that they destroy the inclusions (in a normal atmosphere amber burns and softens but does not liquefy). Exhumation is done by using a very fine saw, viewed under a microscope, to circumscribe a groove around the specimens. When the groove is close enough to the inclusion, the two halves are carefully pried apart, generally separating along the body wall of the insect or plant part. Although only the most common inclusions in amber are used for this work, opening a common stingless bee in Dominican amber brings a sense of mystery, not unlike the unwrapping of an Egyptian mummy. In the case of the bee, however, we can peer into remains that are 25 million years old, not 6,000 years, and the degree of preservation would have inspired even Egyptian morticians.

Ancient Egyptians had to eviscerate and "debrain" cadavers for mummifica-tion, since these organs would bloat and otherwise decompose (the viscera and vital organs were preserved in special canopic jars with spirit fluids like wine).

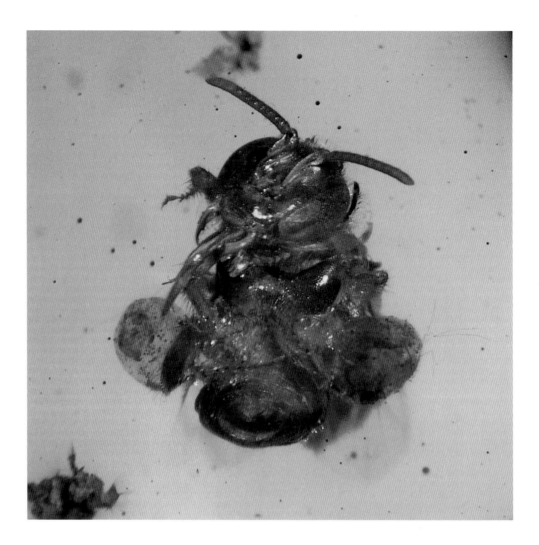

Stingless bee, Proplebeia
dominicana, *the pollen baskets on
its hind legs filled with pollen and
resin that it harvested. These bees
are common in Dominican amber.*
American Museum of Natural
History (Entomology)

Insects in amber, however, often have the digestive tract and brain lying neatly in
place; even the membranous tracheoles, which are the fine tubes that meander
throughout the organs delivering oxygen, remain intact. Insects have no lungs,
but strong fliers (like bees) have air sacs for residual storage of air. The delicate
air-sac membranes in some amber bees lie crumpled, like an opened sheet that
slowly fell to the floor.

A particularly informative lesson in ancient anatomy comes from the muscles
of insects in amber. Insects, in fact, have among the most complicated muscu-
lature of any animals, with minuscule muscles powering, for example, the
movement of the neck, the touch of an antenna, or the flick of a tiny tongue.
In some cases, these muscles are frozen in their original positions. The fine
striations that Kornilowitch described so long ago are there, too. When muscles
contract, filaments of actin and myosin proteins slide past each other. Where
the ends of the filaments align, bands are formed. Under 20,000 times magnification, the bands are obvious. Between the fine bundles of muscle tissue, the
myofibrils, can be seen fingerprint patterns: these are the *mitochondria,* too tiny
for Kornilowitch to have seen. Mitochondria are the "powerhouses" of the cell,
generating most of its energy. Internally, the mitochondria have an intricate

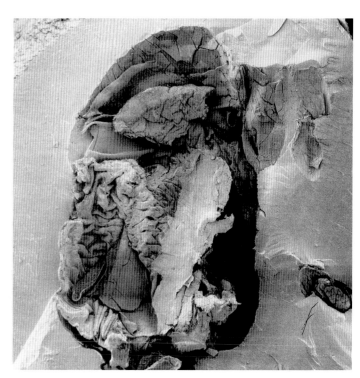

Scanning electronmicrographs of fossils "exhumed" from Dominican amber

Above left: Right half of a stingless bee. Note the sheets of intact muscle in the thorax.
Above right: Left half of bee at left
Below left: Detail from bee above, of the head, showing scales on the tongue, muscles in the head, and the brain
Below right: Detail from bee above, of the thorax, showing crumpled air-sac membranes

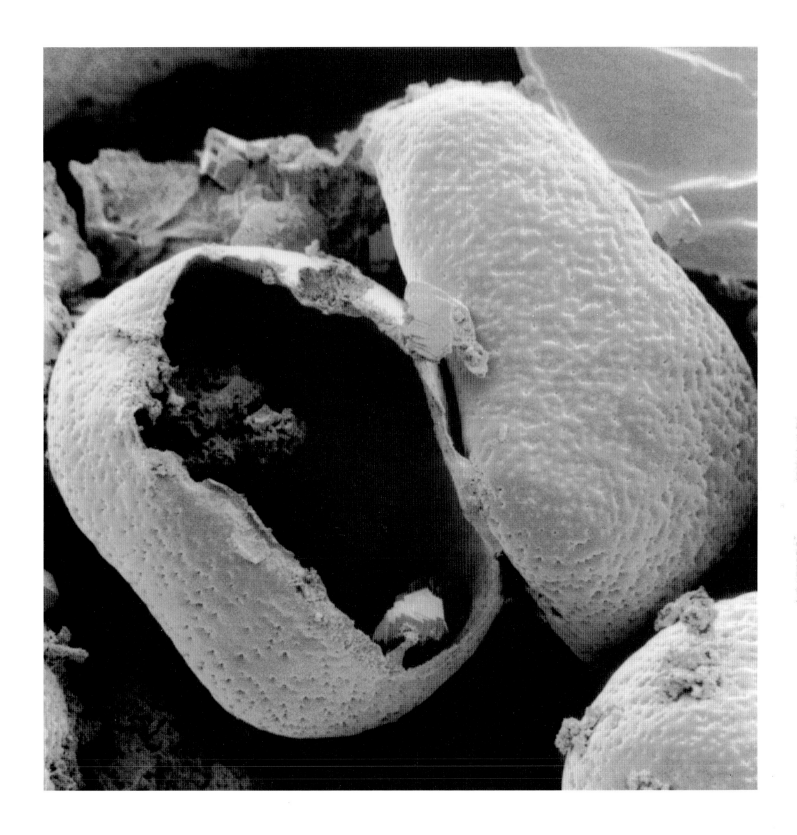

These two pages: Scanning electronmicrographs of fossils "exhumed" from Dominican amber, featuring details from the bee shown on pages 118–19, of a clump of pollen (opposite) and an individual grain (above). The pollen was ingested by the bee.

maze of membranes, which is the fingerprint pattern. Because the muscles in insects must power the wings to beat hundreds of times per second, insect flight-muscle tissue, then as now, has more mitochondria than any other kind of tissue known.

Insects are not the only organisms to be preserved so well in amber. *Anthers* are the club at the apex of a long filament, which produce the pollen in a flower. Anthers of the Dominican amber tree no doubt littered the ground, probably the way those of living *Hymenaea* trees do today, and many of them became stuck and immersed in resin. Exhumed from Dominican amber, anthers show a few pollen grains and a curious carpet of short, fine cells: a pollen grain developed at the end of each one of these cells. The outer coat of pollen (exine) is intricately and geometrically sculptured, with a different pattern for many groups of plants. Exine is also very resistant to decay, to the extent that exines fossilized in sediments are commonly used to study thousands or even millions of years of change. Oddly, exines of pollen in amber are not well preserved. In at least a few cases, though, clumps of pollen attached to stingless bees in Dominican amber were preserved with remarkable fidelity. They were so completely preserved, in fact, that two types of pollen from one bee were discernible, indicating that the bee fed from two kinds of flowers.

Plant tissues preserved in amber have not been extensively studied, but in several cases, small leaves in Dominican amber were exhumed. To the naked eye, the surface of the exhumed leaf is cracked and crumbly, like a dry riverbed. Under high magnification, though, the columnar cells of the epithelium are still neatly stacked. No doubt, if the cells were sectioned into micron-thick slices for closer study, organelles like chloroplasts (the mitochondrial equivalent in plants) would be found.

One of the most curious instances of amber preservation concerns the platypodid and scolytid bark beetles. As discussed earlier, these beetles are also called ambrosia beetles because they inoculate into their wooden galleries a fungus that they feed upon. The fungus is found nowhere else except in these galleries and in specialized pockets in the body of the beetle (called *mycangia*). The species of fungus is specific to the species of beetle. Bark beetles exhumed from Dominican amber show mycangia, replete with spores and filaments of their symbiotic fungus.

An aspect of amber preservation that is just as intriguing as *what* is preserved, is *how* something is preserved. There is often virtually no shrinkage of soft tissues and no traces of decomposition, at least in Dominican and Mexican ambers. (Baltic amber, it has been recently discovered, does not preserve internal tissues as well as Dominican or Mexican amber does, and this is undoubtedly due to its unique chemistry. Baltic amber does show some decomposition and shrinkage, but not always. Chemically and botanically similar fossil and subfossil resins, like copal from East Africa and Colombia, theoretically should show at least similar preservational fidelity, although this has not yet been tested.)

Pages 123–25: Scanning electron-micrographs of fossils "exhumed" from Dominican amber

Above: Small leaf, its surface appearing dried and cracked like a parched lake bed

Below: Epithelial cells of the leaf above, perfectly stacked

*Above: Wood-boring ambrosia beetle,
showing the special pocket for
transporting its symbiotic fungus*

*Opposite: Detail of pocket from above,
still containing the symbiotic fungus*

Preservation of a tiny organism by the original, fluid resin must have been very quick to account for the virtual lack of decomposition. The favored hypothesis is that a very volatile fluid in the resin, perhaps a terpene, seeped rapidly through the body walls and into the tissues, fixing them. Water must have been extracted in this process, since DNA can be naturally preserved only by dehydration. Perhaps the reddish or cloudy halo often found around inclusions in amber are aqueous remains, sequestered by the surrounding resin. But mere dehydration would leave the tissues looking shrunken and shriveled: they must have been fixed, as in embalming.

Antibiotic properties of the resin is another important preservational factor. When a small organism becomes encapsulated by flowing resin, processes of

fixation, dehydration, and sterilization begin immediately. Since the resins harden quickly, a hermetically sealed tomb is soon formed. In a piece of amber in the Stuttgart museum, the branches of a liverwort are preserved with bubbles of liquid. The liverwort extends into a large bubble, where bits of its stem and leaves float free (not unlike the "snow" in one of those snowstorm paperweights). If the water in the bubble were not completely sterile, the bits of plant would be at least partially decomposed.

Whatever the exact mechanism of preservation in amber, at least some resins preserve much finer details and more consistently than any other kind of fossil. In order to showcase such a unique mode of preservation, a special term — "ambalming" — can be coined.

Ancient DNA, Evolution, and Suspended Animation

If thou couldst but speak, little fly, how much more
would we know about the past!

—Immanuel Kant

Biologists who discover, name, describe, and classify various organisms are called systematists, or taxonomists. Perhaps because the science preceded virtually all other branches of biology, or because Victorian taxonomists merely pigeon-holed species, systematics has often been perceived as a science outpaced by biochemistry, physiology, and the like. This was most true of paleontology, where the descriptions of forms in rock required, at best, a microscope. How technical a field is seems to relate directly to how scientific that field is believed to be, regardless of the conceptual bases behind it. Systematics is hardly short in concepts, for it seeks to examine evolutionary relationships among species, and this is at the heart of all biology. Traditionally, it has been done by analyzing the anatomy, or morphology, of an organism, so as to define features that link species, such as six legs defining (in part) the insects.

Technical advances in the 1980s and 1990s then allowed the sequencing of amino acids in proteins and the nucleotides in DNA. DNA is of such interest because it is the molecular basis of inheritance. If a species acquired a mutation, and its descendants inherited the mutation, then all the changes in a segment of DNA could be used to reconstruct the evolutionary history of that lineage. The millions of fungi, worms, mites, insects, and plants are still studied mor-phologically, but the study of their DNA has become a fervent topic. That paleontology, the bastion of morphological study, would become molecular was inconceivable years ago.

Fossils in amber did a great deal to revolutionize the study of ancient DNA. After 1982, when high-magnification, electron-microscopic study of tissue in an amber fossil revealed new detail, scientists began thinking about the possibility of extracting DNA from it. Early attempts at extracting and sequencing DNA from amber fossils were complicated by a cumbersome process of "cloning": taking a segment of DNA, inserting it into the genome of a bacterial colony, and literally growing lots of the DNA, which could then be sequenced. But to start, one needed much more DNA than exists in the pinhead-size piece of tissue one often gets from an insect in amber. It was not until early in 1992 that serious efforts resumed to obtain DNA from fossils in amber.

Probably the event that most sparked interest in ancient DNA from amber was the report in 1990 of intact DNA from 17-million-year-old fossil leaves, not in amber but from clay sediments of Clarkia, Idaho. It had actually been reported years earlier that leaves fossilized at this site contained various preserved complex

Extinct termite, Mastotermes electrodominicus, *in Dominican amber. Length of amber 1.8".* American Museum of Natural History *(Entomology)*

A specimen similar to the one above yielded one of the first DNA sequences recovered from an amber fossil.

biomolecules like chlorophyll (which makes plants green) and other pigments. Even chloroplasts in the cell (which contain the chlorophyll) were preserved. But the pigments degraded rapidly upon exposure to air; some leaves looking originally like colorful fall foliage quickly blackened. Still, these were no ordinary leaf-compression fossils. Two laboratories independently published short sequences of DNA from the same chloroplast gene, one for an extinct magnolia, the other for an extinct bald cypress. The magnolia DNA was very different from living magnolia DNA; in contrast, the bald cypress was very similar to living relatives. Despite the remarkable preservation of these fossil leaves, hundreds of attempts were required for the few successful extractions. A portable extraction laboratory had to be set up at the site to process the specimens before they degraded.

Serendipitous preservation is one part of the Clarkia fossil success; the other part is the technique that was used, Polymerase Chain Reaction (PCR). This technique uses an enzyme from the *Thermophilus aquaticus* bacterium, which lives in hot springs. The enzyme is heated with segments of DNA called *primers* that are specific for a certain gene, the four building-block nucleotides of DNA (A, T, C, and G), and extracted ancient DNA; the mixture is then cooled. The result is large quantities of the ancient DNA from absolutely microscopic amounts, even from a single strand of DNA. Scrupulously clean conditions are required, for the technique is so sensitive that contaminant DNA is also easily amplified. (This caveat is sometimes used to claim that reports of DNA from fossils are not ancient DNA at all, but modern contaminant DNA. There are ways to address this problem; see, for example, page 130.) The PCR technique has revolutionized biology and even forensics. It was this technique that allowed the discovery of the oldest intact DNA yet known, from amber.

Later in 1992, two independent papers simultaneously published DNA from a termite and a bee in the 25-million-year-old amber from the Dominican Republic. Given the consistently fine preservation of this amber and availability of inclusions, these were good first choices. A group from the American Museum of Natural History reported on the DNA from the large primitive termite *Mastotermes electrodominicus*. A group from two universities in California reported the DNA sequences from the common stingless bee *Proplebia dominicana*. (The American Museum of Natural History study had been completed and submitted for publication two months before the completion of the California study, which ultimately was published a month earlier.) In both cases, only a few fragments of DNA, of 200 to 300 nucleotides long, were recovered. The same gene in both cases (called 18s rDNA) was examined, but it was a mere fraction of the entire gene. Some genes may have, say, 10,000 nucleotides, and the entirety of DNA in many complex organisms is housed in 10,000 or more genes. Even though they were such minute fractions of the genomes, the fact that DNA could be preserved this long was instant news, particularly since the reports came in the wake of the book *Jurassic Park* and just prior to the film. What was not so widely popularized was the scientific reason why at least the termite's DNA was of such interest.

Mastotermes electrodominicus is a large, morphologically primitive termite whose only close living relative, *M. darwiniensis* (named for the naturalist), thrives in Australia. A similar extinct species is found in Mexican amber, and compression-fossil *Mastotermes* occur in rocks from North America and Europe. The large termites resemble cockroaches not only in size but also in various features: they have a large pronotal plate that partially shields the head (the heads of roaches are entirely or almost entirely concealed); the eggs of the living species are laid in clumps, like a rudimentary cockroach egg sac; and the wings have numerous veins (in most termites, there are very few veins, and these are very faint). So, at least during Eocene to Miocene times (about 40 to 20 million years ago), the genus was very widespread. Now it is relict, being restricted to the northern, tropical areas of Australia and a few spots in New Guinea (oddly, *M. darwiniensis*

Evolutionary relationships of termites, cockroaches, and praying mantises, based on DNA from living species and from the amber fossil termite. The extinct Mastotermes retains features of its roach ancestry, but it and its living relative are genetically all termite.

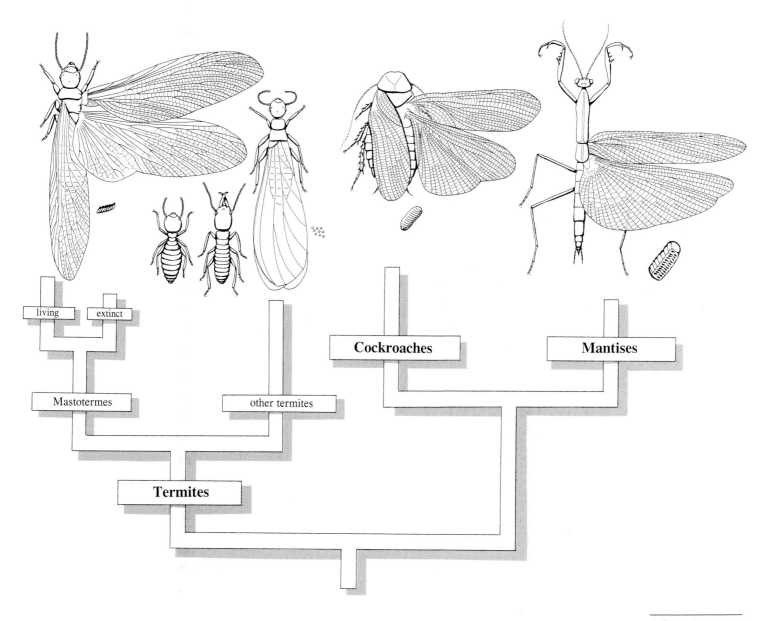

is a serious pest, so why its close relatives became extinct around the world is an enigma). *Mastotermes* could only be considered a relict if all of the fossil and the one living species were closely related. Systematists traditionally grouped those species on the basis of features that also are seen in cockroaches, which evolved well before termites evolved. This would be comparable to defining humans by the presence of hair, which evolved in the earliest mammals, well before humans. Thus, one question was, Are the living and fossil *Mastotermes* indeed closely related, or do they just share a primitive morphological resemblance? The study of DNA would circumvent the morphological problem. Answering this question had implications for evolution, specifically about extinctions.

The other question involved the wider evolutionary relationships of *Mastotermes* to other kinds of termites and to roaches. Because they retain so many roachlike features, *Mastotermes* had often been thought of as "missing links" between the two orders. In this light, termites were seen as highly reduced, myopic, wood-eating, social cockroaches.

DNA from the extinct termite revealed that it and the living species are definitely all termite, not at all "missing links" with cockroaches. An evolutionary tree drawn only with the DNA from the living termites and roaches gave one arrangement. With the addition of DNA from the fossil termite, a slightly different arrangement resulted. Apparently, an evolutionary tree based on living DNA alone can give an incomplete picture, but here was the first glimpse as to how incomplete that picture could be. Other unexpected results appeared in the DNA of the extinct *Mastotermes,* but enough similarities between it and the living species existed to show clearly that the two species truly were closely related. They did not just share a primitive resemblance. Thus, perhaps all fossil *Mastotermes* are closely related. If that is the case, Darwin's *Mastotermes* in Australia, albeit a serious pest, is indeed an evolutionary relict. (Further, the way this DNA matches with its close living relative is a good test that the DNA is authentically ancient.)

In June 1993, newspaper headlines declared, "DNA from the time of dinosaurs discovered." They were based on a report, published in the scientific journal *Nature,* of DNA sequences from a weevil in 120-to-130-million-year-old amber from Lebanon. The report was from virtually the same California group that earlier had announced DNA from Dominican amber bees. Curiously, the scientific study was published on the day the film version of *Jurassic Park* opened. The DNA had been taken from a single, small nemonychid weevil in the amber. Today, nemonychids feed mostly on conifers, even on araucarians, so it was probably feeding on the tree that produced the Lebanese amber. The scientists examined two fragments of the same gene (18s rDNA) that was studied earlier for the termite. As expected, the fossil DNA was quite primitive. What fueled publicity was that DNA this ancient, from the age of dinosaurs, could exist. Of course, the *Jurassic Park* premise, of cloning extinct creatures from these snippets of DNA, seemed all the more compelling.

Since then, DNA has been recovered, in five different laboratories, from leaves, fruit flies, wood gnats, a lizard, leaf beetles, and a fungus gnat, all in Dominican amber. Approximately one in every three attempts is successful. Such incredible consistency of DNA preservation must be attributable to the desiccating properties of resin. No doubt, many more DNA extractions will be made from amber fossils, which has implications not only for evolutionary studies but also for scientific ethics.

Suspended Animation? The DNA that is preserved in amber fossils is so hopelessly jumbled into tiny fragments that, given present technology, it would be impossible to reconstruct the entire genome of, say, an extinct termite. The task has been likened to the scale of reconstructing Tolstoy's *War and Peace* out of the alphabet noodles in soup, without ever having read the book. Even with an almost limitless supply of tissue and fully intact DNA, the human genome project is still requiring decades of work by hundreds of researchers. If, in fifty years, the technology exists to read and assemble the genome of our extinct termite, could the termite then be resurrected? This is actually the most complicated part, since so many complex levels of organization lie between the DNA sequence and the whole organism. How the DNA

Bubbles in Dominican amber, harbor of quiescent, ancient bacteria? Length of amber 1.8". Private collection

lies in a scaffolding of proteins on the chromosomes; how the thousands of genes turn off and on, and when, are just some of the questions. Even if all this were possible, hundreds of years from now, the ultimate question is, Should it be done? Presumably by then, the concern would not be so much with resurrecting extinct creatures but with human immortality. In the meantime, there are other concerns.

The remarkable preservation in amber has given at least one scientist a vision of "life in amber." If DNA is preserved, why couldn't simple organisms like viruses, bacteria, protozoa, and fungi also be preserved? At least some forms of these have spores and cysts that remain dormant and extremely long-lived under the desiccated conditions amber provides.

In May 1995, the apparent revival of a *Bacillus* bacterium was reported to have been extracted and cultured from the common stingless bee in Dominican amber. Excellent precautions were taken to insure against contamination, and the DNA of the bacterium was very similar, but not identical, to another kind of *Bacillus* known to live today in bees. If the results prove true, how we view the mortality of organisms needs to be revised. Widespread skepticism exists in the scientific community, though, as to whether this bacterium is indeed ancient.

One problem with trying to determine if the bacteria apparently revived from amber are authentic is that the living flora of bacteria is so poorly known that one may never be sure if a positive result is simply due to some unknown modern species contaminating the culture. In a teaspoon of forest soil thrive thousands of species of bacteria, most new to science. What assurance is there, given the most sterile and careful conditions of isolation, that a weird bacterium is authentically ancient? Also, all of the DNA extracted thus far from organisms trapped in amber is extremely fragmented. Given this, how is it posssible that an entire genome (the DNA chain in an organism) can remain entirely unbroken? A bacterium with a fragmented genome would never be viable. This work raises questions about the ethics of culturing extinct microorganisms. Some laboratories are even trying to insert fragments of the (presumably) extinct bacterial DNA into close living relatives. This raises even more concern than the current one over recombinant-DNA research, since it involves constructed microbes whose pathogenic potential is unknown.

Less esoteric and more germane to the lover of amber is the responsibility of scientists as stewards of great collections. If the extraction of tissues from amber, DNA, microbes, or whatever, becomes commonplace, what guidelines and safeguards are there to prevent unique and rare pieces from being harmed or even destroyed? This concern became most pertinent as a result of the study on the Lebanese amber weevil. Extraction of the tissues largely destroyed the specimen (the pieces were then glued back together), which is all the more unfortunate because it was unique. Experiments like this, regardless of how promising the result may be, should be done only on fossils that are common. If any aspect of a fossil is to remain available for study, it should be the morphology—all of it.

PROCESSED AMBER, IMITATIONS, AND FORGERIES

*T*he virtues of amber are many, but two main limitations are that it gradually deteriorates when exposed to heat and air, and forgeries are easily and sometimes convincingly made. Various substances have been used to imitate amber in decorative objects, but these are not usually sold with the intention of deception. Generally, the imitation materials and processed amber are quite easy to distinguish from true, unadulterated amber. Some of the imitation substances include cellulose acetate and nitrate, acrylic resins, Bakelites (the first synthetics used), and now most commonly, polyester resin. Even horn and hardened casein (the protein in milk) have been used as amber imitations. The imitations are usually discerned by their unnatural color or composition, or with a hot needle. When a hot needle is touched to amber or copal, the smell is resinous; imitations will smell acrid, like burning plastic or burned fruit (for the celluloids).

A common process for amber jewelry is the creation of "sun spangles," which are disks at different angles to each other. The disks are made by heating a piece of amber with numerous air bubbles buried in cans of hot sand. Popping heard from the heated can indicates that the bubbles have expanded, creating a discoidal fracture around them like the rings around the planet Saturn. All of the substance is natural, but it has been processed to give the desired effect of these highly reflective disks. This technique is closely related to *clarification*. Most amber pieces used for decorative objects, and particularly for the flat pieces glued into mosaics and onto chests, were clarified. By heating amber in oil very gradually, the minute bubbles near the surface become filled with the oil, and the amber becomes somewhat transparent, even in bony and bastard amber. Despite what some may claim, this technique cannot make a piece of bony or bastard amber completely clear, but it does give a characteristic transparency to the surface. In the time of Pliny the Elder, the oil that was used for clarification was the rendered fat of a suckling pig. Now rapeseed oil is used, partly because its refractive index closely matches that of amber.

An extremely important technique for processing amber has been the pressing of amber to create *ambroid*. Here, the masses of small, unusable chips from amber mining are fused in a vacuum with steam of at least 400° F. The softened amber is then pressed through a sieve, mixed together, and hardened into blocks. It can be dyed, usually a dark red. Ambroid is characterized by its flow lines, which become more apparent with age. The technique was particularly useful for the mass production of standard small objects, such as

Brooch made with a cabochon of amber having "sun spangles," set in silver. This modern piece from Poland exemplifies a typical use for this kind of amber. The amber is produced by careful heating, which causes internal bubbles to expand, creating the disks. Height 1.5". Private collection

Below: Ambroid, or "pressed" amber, made from Baltic amber. These pieces were intended for use in buttons and as mouthpieces for pipes. Length of longest piece 1.4". American Museum of Natural History

Opposite, above: Forgery with a lizard, made from cast kauri-gum resin. Length of amber 6.7". Private collection

Opposite, below: Forgery with a paper wasp in polyester resin, sold in the Dominican Republic. Length of amber 1.3". Private collection

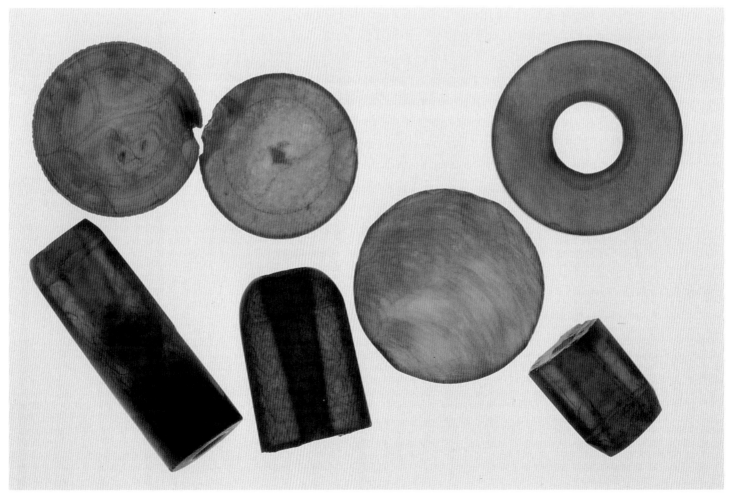

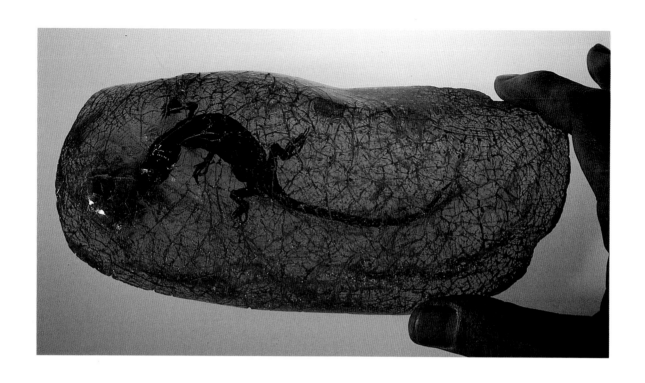

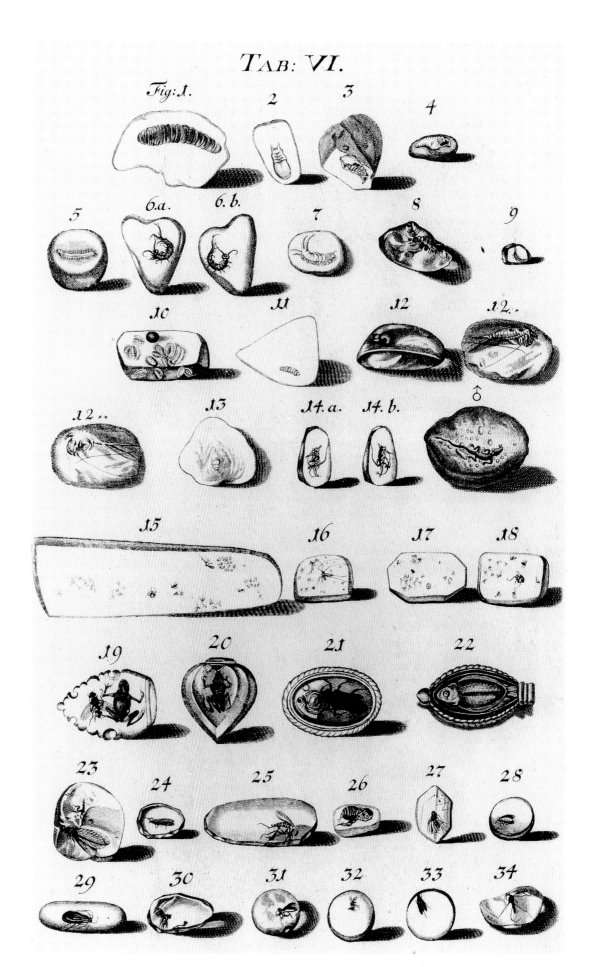

Plate from Nathanael Sendelio's Historia Succinorum Corpora aliena involventium et Naturae Opere..., *1742. Several illustrations depict obvious forgeries; see, for example, the lizard marked ♂ and numbers 19–22.*

buttons and mouthpieces for pipes. At the turn of the twentieth century, the use of pressed amber for smoking implements was very popular, because amber was considered healthier than horn, bone, or ivory, and it was smoother. Small inclusions, such as insects and plant parts, can be embedded in the pressed ambroid while it is soft, but the heat and pressure greatly distort them, and the flow lines are always an indication that the inclusions are unnatural.

The real interest in amber forgeries involves the inclusions of small organisms purposely embedded in an imitation substance or in a cavity of natural amber. Currently, and presumably hundreds of years ago, the most sought and therefore most expensive amber objects contained large insects and small vertebrates. Forgeries of these have been made for at least six hundred years. Probably all forgers deceive their buyers simply for the money. For the few cases in which scientists have been fooled by faked inclusions, the forger almost certainly did not intend to be misleading scientifically: the fake just ended up in a museum collection.

Since it is not feasible to melt authentic amber to insert inclusions, some convincing substitutes must be used. The first forgeries were made from melted copal, a practice common up to the early twentieth century. Even Pliny doubted the authenticity of lizards in Baltic amber; he believed them to be copal forgeries. East African copal (the kind originally used) would be pulverized and mixed with spirits, such as turpentine or alcohol, and then heated until the copal was dissolved. As the solvent evaporated and the mixture became thicker, it could then be poured into a mold, into which was placed, say, a large beetle, scorpion, or lizard. At the peak of the kauri-gum trade in the late nineteenth century, a popular sideline was the manufacture of fossil forgeries, many still in private collections. Most of these forgeries are very unsophisticated, consisting of unnaturally large arthropods and lizards, and usually with their appendages far too neatly arranged. At the very least, copal forgeries are easily spotted.

The caveat is that not all inclusions in copal, including lizards, are forgeries. Some, such as ones in the Natural History Museum in London, misled a few Victorian taxonomists who thought they were in true amber. The American Museum of Natural History has an extensive collection of insects naturally preserved in East African copal. The Brooklyn Children's Museum has a rectangular block of copal containing a small gecko lizard. The immediate assumption was that it is a fake, but close examination revealed microscopic plant hairs and frail midges that a forger would never have thought to include; or, if he had, heating the copal mixture would have greatly distorted them. The piece was probably cut and trimmed from a large chunk of kauri gum, in which a lizard had been naturally trapped (or pushed!). One ironclad technique of discerning a copal forgery relies on some sophisticated chemistry. Copal forgeries, even one made a hundred years ago, contain traces of solvent that have not evaporated, but to detect this requires expensive instrumentation.

Forgeries in polyester resin have largely replaced copal forgeries, perhaps because the technique is easier and, to the inexperienced eye, more convincing.

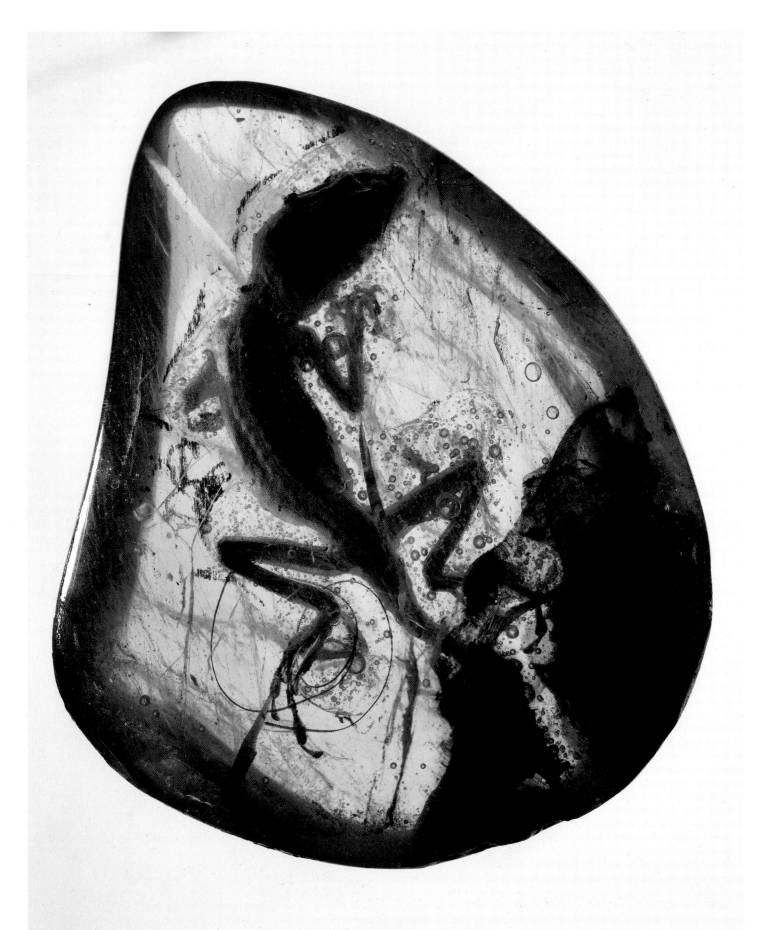

Opposite: Forgery showing a small anole lizard set in a niche carved from Dominican amber. The niche was then filled with polyester casting resin. Note the curled hair and the pen outline around the lizard; otherwise, it would have been a very convincing forgery. Private collection

Two skillful forgeries in Baltic amber, both made by inserting an inclusion in a cavity in natural Baltic amber

Above: Small frog and scallop shell embedded in a central hole bored into the piece. Length of amber 1.9″. American Museum of Natural History (Entomology)

Below: "Piltdown Fly," forgery with the common latrine fly, Fannia scalaris, studied by the eminent entomologist Willi Hennig. The Natural History Museum, London

Since about 1980, polyester forgeries have become very popular in the Dominican Republic and Mexico. Unlike forgeries in copal, detecting fakes in polyester resin is unequivocal. But, like copal forgeries, the inclusions in polyester forgeries, generally lizards, are usually very unnatural. This is convenient, since small vertebrates and the like draw immediate scrutiny by dealers and sophisticated buyers; it is the thousands of small inclusions that do not. Forgers also unwittingly use large, common insects (like modern honeybees) that entomologists easily spot as fakes, but which a layperson may not.

One cannot rely on detecting a forgery based on the extreme clarity of the substance, for some amber, especially Mexican and Dominican amber, can be remarkably free of debris. In fact, forgers of Dominican and Mexican "amber" generally sprinkle dirt into the resin, but to an amber specialist even the dirt is suspiciously unnatural. Like an experienced bank teller who can detect counterfeit currency by the texture of the paper, an experienced amber expert can also detect a polyester forgery by the feel (and color) of the piece. The easiest and surest method of detecting polyester forgeries for laypeople is the hot-needle test.

The most clever technique in amber-fossil forgeries is one of the oldest as well. For at least four hundred years, Europeans have been making fossil forgeries from Baltic amber, and the many natural flow lines and cracks in it lend themselves wonderfully to a special technique. A piece is split along the plane of a crack or fissure running through the amber. A small cavity is carved out of a split surface, filled with the forged inclusion and with resin, balsam, or melted copal, and the same is used for fusing the two halves back together. The difference between the resin and amber is generally so slight as to be undetectable, especially if other flow lines and fissures help to conceal the edges of the cavity. All of the most sophisticated chemical tests would reveal the piece to be authentic amber.

The technique is so clever that it easily misleads even experts. One case involved a piece in the American Museum of Natural History. It was about two inches long, dark red, and with a crazed surface typical of old Baltic amber pieces. Inside was a stream of resin, and inside of that was a tiny frog and what appeared to be two large bubbles. The piece came from one of the most venerable private collections of minerals ever assembled (by Clarence Bement), and the whole lot was purchased for the American Museum of Natural History by J. P. Morgan in 1900, for $100,000. Elation at the rediscovery of "Bement's frog" quickly subsided when the piece was examined closely: one of the small "bubbles" was actually a small scallop, which could not possibly have been caught in amber. It was then discovered that the internal stream of resin was actually a bore hole filled with modern resin and that a slice had been taken off one end to make the hole and then very carefully fused back onto the original piece. One surface of the piece was left naturally rough to help conceal traces of the forgery. Fortunately, the piece was detected before there could be a scientifically embarrassing report on it.

Another case was not so fortunate. In the paleontological collections of the Natural History Museum, London, was a small piece of amber containing what looks like a small housefly. It had been acquired by the museum in the nineteenth century from a distinguished German scientist. More than seventy years later, another distinguished German scientist, Willi Hennig, studied it and reported that the specimen was a fly indistinguishable from the common sewer fly, *Fannia scalaris* (there are at least a hundred other species of *Fannia*, but none are as common). He considered the possibility that the specimen was a forgery, but there seemed to be no signs of it, and so he dwelled, in his original paper, on how an insect species can persist in evolutionary time. It became a cited example of evolutionary stasis, until 1993, when a researcher at the British Museum discovered it was a forgery. Were it not for the heat of a microscope lamp, which caused the piece to crack in a certain way, the specimen might never have come under suspicion. It was such a good forgery that it had stumped even a scientist like Hennig, who was renowned for his work on insects, including flies in amber. It was made in a way similar to that of Bement's frog, by embedding a fly (probably from a windowsill) into a cavity in natural Baltic amber. In allusion to a famous human-fossil forgery, the fake fly has been euphemistically dubbed "Piltdown Fly"; in this case, however, the forger almost certainly never intended to mislead a scientist.

Amber

in

Art

The use of amber as an artistic medium is a natural one because it can be carved relatively easily, and it comes in a variety of warm colors. However, it splinters and breaks more easily than ivory, one substance used often with amber, which is why the detail in amber carvings is not as intricate as that in ivories. The smooth feel of highly polished amber and its warmth make it especially prized for objects that are handled or worn against the skin, such as chess pieces, rosary beads, and necklaces. The pianist Frederic Chopin carried amber pieces with him, to relax his fingers over them before a performance, and the conductor Leonard Bernstein (whose last name means "amber" in German) had a conducting baton with an amber handle. So cherished was amber among the Chinese and ancient Italic-speaking peoples that their carvings would follow the natural contours of the piece. They considered it wasteful to trim off excess amber simply for the sake of making a symmetrical carving. In some cases, this technique would distort a bust or figure, while in others the topography of the piece dictated the actual subject, often yielding a very pleasing asymmetry.

This conservative fitting of a carved subject to the natural contours of a piece was completely opposite to the practices in seventeenth- and eighteenth-century amber workshops of northern Europe, where symmetry was the essence. Large, elaborate artworks were crafted by using many smaller amber pieces in mosaics or as small tiles, often attached to a wooden frame. The stems of some chalices and candlesticks were made from ten or more pieces of amber, all intricately turned and held together by a central, internal rod. The most ornate and contrived artistic application of amber can be seen in the small altars and chests made in an architectural motif. In many cases, a mosaic of amber pieces would be attached as a thin veneer to the wooden frame. Unfortunately, oxidation is a serious problem with wafer-thin pieces, and such seventeenth- and eighteenth-century objects are generally in worse condition than much older pieces made of solid amber. With many of the ancient ambers, too, the conditions of their preservation, in damp, cool tombs, was ideal.

MESOLITHIC PERIOD TO THE BRONZE AGE

For at least ten millennia, European peoples have adorned themselves with Baltic amber. The oldest artifacts are beads and amulets found close to where the most extensive amber deposits occur even today. One can imagine that in an area where there are long winter nights, the sun was a focal point of existence and worship. A substance of such warm color and feel, like amber, probably had special significance to the early Baltic peoples. Amber was such an important commodity with which to trade for copper and iron that it had a fundamental influence on the development of northern European cultures.

The oldest amber artifacts, however, are from England, not the Baltic. Rough beads from Gough's Cave in Cheddar and Cresswell Crags, Starr Carr, dated from 11,000 to 9000 B.C. (Paleolithic), were undoubtedly crafted from pieces of Baltic amber that washed onto the eastern shores of Britain. (This was a time deep into the last Ice Age, when the British Isles were still largely connected to the European mainland.) Similar artifacts must have been made by people living near the Baltic Sea at the same time. Mesolithic (c. 4000 B.C.) amulets and beads are known from southern Scandinavia but are considerably rarer from the eastern Baltic region. Large-scale production of amber artifacts was not evident until the Early Neolithic (3400–3100 B.C.) Narva Culture. A find near Sarnate in the eastern Baltic uncovered amber pieces with flint and bone tools, which were probably used to work the amber. By the Middle Neolithic (3100–2500 B.C.), amber working and trade had burgeoned, especially in the eastern Baltic. In Sventoji, Lithuania, 850 amber ornaments from this period were found, some of them V-shaped buttons of bone, jet, and amber. At Woldenberg, near Berlin, the elegant amber carving of a highly stylized horse from this period was found. The Lake Lubanas region, Latvia, represents the richest source of Middle Neolithic amber workings. Enough pieces were crafted for widespread trading throughout Europe. Beads are known from the Middle Neolithic of Charavines, in the alpine region of France. Perhaps the most distinctive type of artifact in amber is the double-headed ax bead, most common from southern Scandinavia and Mecklenberg.

The most celebrated of all Neolithic amber finds was made in the 1850s near Juodkrante (then Schwarzort), Samland Peninsula. Dredged from the Kurisches Haff were 434 artifacts, including beads, buttons, small human "idols," animals, and ax-shaped beads with holes for suspending as pendants. They were deposited in the collection of Albertus Universität, Königsberg, but, like the university's

Pendants from the Mesolithic period of western Jutland, Denmark. Oval pendant decorated with lines of drilled holes, Maglemose Culture, c. 7000 B.C. Height 2.1". Stylized bear, probably Ertebolle Culture, c. 4500 B.C. Length 2.6". Nationalmuseet, Copenhagen, A52125 (oval), A52089 (bear)

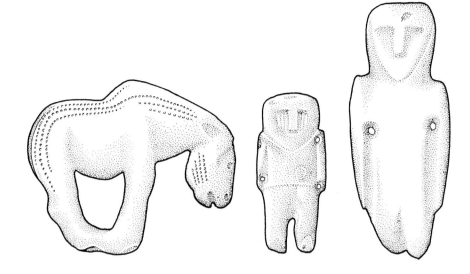

Horse, from Woldenberg (present-day Dobiegniew, Poland), c. 2000 B.C. Length approximately 4". Human figures, from Juodkrante (old Schwarzort), Lithuania, c. 2000 B.C. Heights 2.6", 5.6". Collection Palanga Museum, Lithuania

Among the earliest artifacts in amber, from the Mesolithic of northern Europe, these pieces were probably sewn onto a garment through the holes.

huge collection of fossiliferous amber, the archaeological collections were dispersed to other areas of Germany during World War II. Similar artifacts have been found in Palanga, Lithuania, in situ in peat bogs or grave sites. Those pieces are now in the Palanga Museum, which is largely dedicated to amber.

During the Neolithic period and the much later Iron Age, the British Isles are rather peripheral to our understanding of amber in its archaeological context. This is not true, however, for the Bronze Age (3000–700 B.C.), of which more is known in Great Britain than in continental Europe. Documentation exists from

Reconstruction of an Upper Paleolithic homesite in the Ukraine. American Museum of Natural History (Hall of Human Biology and Evolution)

Similar peoples crafted the earliest known artifacts from Baltic amber, or even from amber collected more locally, from the Dnieper Basin. The warm color and feel of amber must have given it a very special significance.

about A.D. 1500 of amber pieces stranded on the eastern shores of England and Scotland. Judging from its succinic-acid content, the material originates from the Baltic Sea. One estimate is that nine to twelve pounds of amber are stranded each year in the British Isles; there are even reports of pieces weighing two pounds. Thus, the raw material for Bronze Age amber artifacts was available locally; Continental trade was not necessary but probably occurred. The early Bronze Age (Bell Beaker phase, or Wessex Culture) saw some remarkable developments, with ambers preserved in situations of special social significance.

Considered by some to be the greatest archaeological finds of amber are two early Bronze Age amber cups, one from Clandon Barrow, the other from Hove. The latter, dated at 1285–1193 B.C., was found in 1821 in the grave of a tumulus mound fifteen to twenty feet high and forty feet long at Hove, near Brighton, England. The cup was buried with the partially cremated remains of a man laid in a hewn log coffin. With the cup was a "celt" (battle-ax), whetstone pendant, and a bronze dagger. The cup has a crazed surface but otherwise is in excellent condition. Made from a piece of clear red amber, it was carved and polished with unexpectedly fine artisanship and symmetry. It is small (three and a half inches in diameter), with a handle and eight fine, carved rings encircling the top. The Hove tumulus cup may have been used in ceremonies, perhaps by the ruler with whom it was buried.

The Wessex Culture was a hierarchical, male-dominated society in which amber was particularly cherished. Women were buried with necklaces of amber

Hove tumulus cup. Diameter 3.5".
Booth Museum, Brighton, England

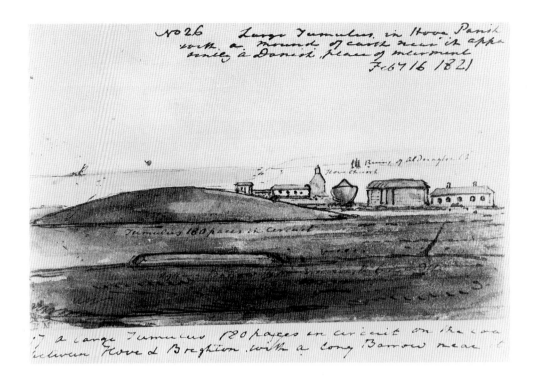

Field sketch and written notes of the Bronze Age Hove tumulus mound, where the famous amber cup, perhaps the finest archaeological artifact of amber, was unearthed

beads, and important men with special artifacts of amber. Archaeologists have uncovered amber artifacts even among the ruins of Stonehenge.

Wessex Culture amber beads may be extremely important in revealing a link between the British Isles and Mycenae during the Bronze Age. In particular, spacer plates, used to keep several strands separated in necklaces, with a distinctive pattern of borings have been found in both Mycenaean and Wessex burials. Did the Mycenaeans actually import amber beads from the Wessex Culture?

A similar exchange took place during continental Europe's Iron Age. This was especially the case during the Hallstatt period, 700–400 B.C., when the Hallstatt Culture traded most extensively with the Etruscans. Some Etruscan amber artifacts found their way back north, in fact, in burials of ruling men from what is now the border of Yugoslavia and Bosnia and Herzegovina. Hundreds of amber beads were included in many burials of the time, particularly those of women. It was during this period that the famous ancient amber routes were established, creating a major north-south axis in amber trade that linked the Baltic with the northern Mediterranean. The routes began at the Jutland coast or the Samland Peninsula, moved down the Elbe River near Hamburg, down the Danube, through the Brenner Pass over the Alps, to Aquileia and centers at the mouth of the Po Valley. From here, amber made its way to various centers in central and southern Italy and around the Mediterranean.

AMBER AMONG THE ANCIENTS

Him the thunderer hurled
From the empyrean headlong to the gulf
Of the half-parched Eridanus, where weep
Even now the sister trees their amber tears
O'er Phaëthon, untimely dead.

—John Milton

Such is Milton's brief version of one Greek legend on the origin of amber. Phaëthon, son of the sun god Helios and of Clymene, asked to drive a chariot across the sky. Helios warned him not to whip the fiery horses, but Phaëthon did this anyway. The horses bolted and came too close to Earth, resulting in a drought. Earth complained to Zeus, who struck Phaëthon dead with a thunderbolt, and his body fell into the Eridanus. Phaëthon's sisters, the Heliades, and his mother, Clymene, wept over the body so much that they became rooted where they stood, their clothes turned to bark and their bodies transformed into poplar trees. The tears that dropped into the Eridanus hardened into drops of amber.

Gaius Plinius Secundus, or Pliny the Elder, the Roman scholar to whom we owe most of our knowledge about ancient views of the natural world, scorned the "frivolities" of amber, both its use in Roman adornment and the absurd Greek legends about its origin: He discounted Sophocles' tale that amber was produced in countries beyond India, from the tears that are shed for Meleager by the birds called meleagrides, and the legend of Plutarch, that amber was formed from the urine of the lynx, "lyncurium." Ironically, Pliny himself propagated several myths about amber, one being that it came from India and Syria, where no appreciable quantities exist. He might have been referring to large deposits mined in Burma (Myanmar) since the first century A.D., but his reference to pieces with lizards is almost certainly of East African copal. In chapters 11 and 12 of Book 37 of his *Natural History,* he also stated that "undoubtedly [amber] is a product of the islands of the northern ocean, a substance called *glaesum* by the Germans." (*Glaesum* derived from *glaes,* or "glass"; one island in this northern ocean, the Baltic, the Romans even called Glaesaria.) Pliny conjectured that the actual river represented by the mythical Eridanus in the legend of Phaëton was the Po in northern Italy (called Padus in Pliny's time). But no natural amber deposits have ever been found in this region, and confusion may stem from the geographical importance of Spina, at the head of the Po, which was a vital fifth-

Winged female head. Etruscan, 5th century B.C. Height 3.1". Private collection

Carved from an amber piece that was originally a deep, transparent red, the surface of this, the largest and finest ancient head sculpted from amber, is now crazed and opaque but consolidated. The head exhibits hallmark features of Etruscan portraits from this period: large eyes with the upper and lower lids well defined, but no brow; straight nose and small mouth and chin. The hair is rendered in waves of parallel grooves across the forehead, and the ears are small and featureless. The wings, with eight feathers in each, have a thin base along the neck behind the ears. Between the wings and above the hair is uncarved (but polished) amber showing natural contours and pitting. The piece is precisely depicted and probably represents a divinity such as Demeter or Kore.

Opposite: Winged deity with a youth. Etruscan, c. fifth century B.C. Height 4.5". Private collection

This work was carved from a homogenously opaque, reddish-brown piece of amber; a consolidant appears to have been used, making the surface slightly more shiny than it would be naturally. The hair of the deity is pulled back with a band and falls in fine braids down the back. The right wing is folded down under the right arm, with the right hand resting on the left thigh of the youth and the left hand wrapped behind his neck. The youth is distinguished from the deity by his relatively larger eyes, finer features, and smaller size. His right arm is raised, and the left arm rests on his lap. The youth appears to be sitting on the lap of the deity. The group may represent Eos (in Etruscan, Thesan), the Greek goddess of the dawn, carrying off one of her young lovers, perhaps Tithonos (Etruscan, Tinthun).

The influence of the natural contours of the amber is evident, particularly in the margins around the youth's raised hand and the right side of the piece, for which no actual structure is rendered. The bulging chest of the deity and the depression in the breast of the youth reflect deformities in the amber. The large hole between the figures was bored, perhaps for suspending as a pendant, but the one in front of the deity's chest appears natural.

to-second-century B.C. Etruscan trading center for amber coming from the north. Others speculate that the mythical Eridanus was a northern European river, such as the Elbe, near the Baltic coast, but the ancients had very little direct knowledge of this area until the first century A.D.

What Pliny did not mention was that even Aristotle recognized that the inclusions in amber indicated a liquid origin, that amber was "petrified poplar gum." More detailed ancient knowledge about Baltic amber comes from the *De Germania* (A.D. 98), written by the Roman historian Cornelius Tacitus. Tacitus was the first to write that beyond the land of the Goths lay the Aestii people, who gathered amber (*metallum sudaticum,* or "exuded metal") that had been washed up by the sea. They sell it, he said, because "Roman luxury gives it repute." Tacitus was also the first to conjecture that the amber forests were remote from where the amber was found, although he believed that the amber and its trees were contemporaneous.

Of great interest is that Diodorus, one of the many Greeks who had settled in Sicily, made no mention of Sicilian amber: it probably was not being harvested during ancient Greek, Italic, and Roman times. The Baltic coast was virtually the sole source for the amber used for artifacts throughout Europe and even Asia Minor, evidenced by Roman artifacts along the amber trade routes as well as by examination of the distinctively high succinic-acid content of the ancient amber pieces.

Pendant in the form of a sphinx. Italic, 600–550 B.C. Length 2.8". Private collection, New York

The head of the crouching sphinx is folded back, showing a profile, and the legs are folded beneath the body; a tail is curled up over the back, the mane is smooth, and the left wing is visible. The piece originally was worn as a pendant, with the body vertical and the head horizontal. Sphinxes like this one, the largest and finest Italic amber sphinx surviving from antiquity, would have served as an amulet to ward off evil in this and the afterlife.

Pendant of amber and gold. Etruscan, 7th century B.C. Maximum width 2.25". Private collection

A small resting dog carved from amber is incorporated into a pendant made from a thin, gold setting that swivels on an oval loop, with a small tube at the top for suspending it. This pendant is unique; all other similarly shaped pendants are of silver and without figured decoration. Without back-lighting, the amber appears opaque because of the surface oxidation (above); its natural translucency is revealed by strong backlighting (below). Both bezel and loop are embellished with granulation.

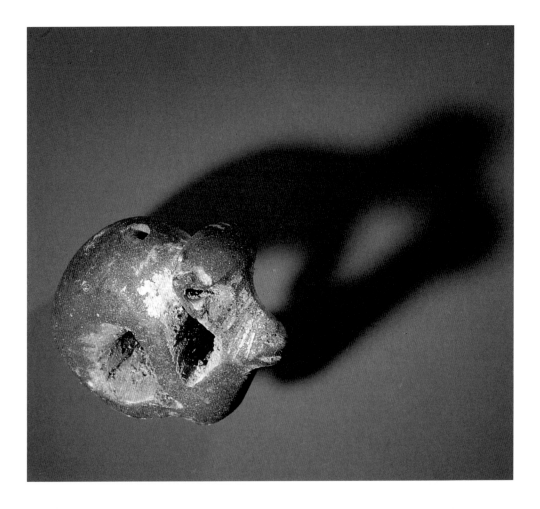

Pendant in the form of a monkey. Etruscan, seventh century B.C. Height 1". Collection Mr. and Mrs. Robert Haber, New York

As early as the Egyptian Old Kingdom, monkey amulets were sexual charms in this life and the next. For the Etruscans, monkeys were a symbol of love and sexual fulfillment in the afterlife. This monkey was carved in a typical pose, with elbows resting on the knees and the long snout resting in the hands. Shallow parallel grooves were inscribed into the snout. It was strung through holes representing the ears. Monkey pendants of this type have been found in Vetulonia, Narce, and Veii.

Finger ring. Roman, second century A.D. Largest diameter 1.4". Private collection

The ring was carved from a single piece of clear red amber; its surface is extensively crazed. A small oval plaque of carnelian inscribed with the profile of an eagle (a popular intaglio device from this period) is situated on the widest part of the ring. Amber rings were popular between the reigns of Nero and Septimius Severus.

By the Middle Bronze Age, the "gold of the North" reached the Mediterranean, where it was an important burial item. It was worked into beads and into more elaborate necklaces; one Shaft Grave tomb at Mycenae contained hundreds of such beads. Other finds from elsewhere in Greece, Crete, the Ionian islands, Palestine, and Egypt indicate widespread trade of the material from the north (a recent study even suggests that a number of objects from the tomb of Tutankhamen are of amber). By the end of the Bronze Age, however, little amber was reaching the Mediterranean. It was not until nearly five hundred years later that amber was to become popular or once again available in large quantities.

During the Iron Age in Greece and Italy, amber was used extensively in jewelry and decorative objects. Throughout the Italian peninsula, huge amounts of the material were carved into small figures and beads and used to decorate pins. Some pieces were of enormous size—one seventh-century B.C. fibula decoration is documented as weighing more than two pounds. Among the Etruscans and the early Latins, amber was carved into jewelry elements and into amuletic images, including monkeys, horses, nude females, and heraldic compositions of humans and animals.

Many of the early Italic ambers are Etruscan, perhaps because amber preserved so well in dark, damp tombs, where it accompanied the dead among the extensive Etruscan necropolises. The civilization of Etruria flourished between approximately 900 and 200 B.C., in the area of Italy centered around Tuscany. Etruria was more a league of city-states than a nation, bound by common rituals and a distinctive, still largely undeciphered language. Etruscans were a people of contrasts: pastoral but with some taste for brutal sports and human sacrifice; extensive traders who also had a greatly developed agriculture and system of animal husbandry. The Etruscans exported iron and imported slaves and furs from the Celts; pottery from Corinth; ostrich eggs, gold, and ivory from Carthage; and amber from the Balts. The trading ship *Giglio,* which sank off the west coast of Italy about 600 B.C., was carrying metals, pottery, flutes, and amber. Etruscan women held unusually high social status for the period, and they were the chief users of amber objects.

After about 600 B.C., very little amber is documented in Greece or Italy. For the following three centuries, including the classical period in Greece, the archaeological record is nearly blank for both the Greeks and the Etruscans. In contrast, outside of the Etruscan areas of Italy, particularly in the south of the Italian peninsula, objects excavated from tombs show that amber was an important grave furnishing. The thousands of beads, figured pendants, and pin decorations that have come to light in modern times were produced by many of the peoples in Italy: some by the Etruscans, others by indigenous Italics influenced by Etruscan and Greek art, and still others were made in Campania and in the Greek colonies of southern Italy. The majority appear to have been worked by the Italic peoples. Favorite subjects for the often massive pendants are derived from the ancient Mediterranean canon of magical subjects, including female heads, mythological creatures such as satyrs, sirens, and sphinxes, and real creatures of African genre: scarabs, monkeys, lions, and gazelles.

The low-relief carving technique is standard for the facial features of the heads: large, almond-shaped eyes, small nose and mouth, and hair rendered by shallow, parallel grooves. Etruscans were also accomplished goldsmiths; they developed a distinctive granulation technique in their jewelry. The jewelry shows an influence of the western Greeks but with a more ornate style. In some cases, small carvings of amber were set into silver or gold pendants. Two necklaces and pairs of earrings made in replica of Etruscan goldwork are in the Arnold Buffum collection at the Museum of Fine Arts, Boston. Each pendant and earring is a cabochon of Sicilian amber set in gold, although the pieces themselves are not known to be modeled after an original object.

During the height of the Roman Empire, long-distance trading flourished. Prized substances included ivory from Ethiopia, frankincense from southern Arabia, pepper from India, silk from China, and flaxen-haired slaves and amber from the Baltic region. About A.D. 54 to 60, Nero dispatched a Roman officer to find the source of amber; he reached the Baltic coast and returned with hundreds of pounds of the substance, some pieces weighing several pounds. Roman coins from A.D. 138 to 180 scattered throughout the Gulf of Danzig testify to the Roman occupation of the Amber Coast. So charmed were the Romans by amber that, according to Pliny, they sometimes prized a figurine carved from a piece of amber more than a slave. One can only imagine the cost of the amber portrait, perhaps a bust, of Augustus at Olympia, which was reported by the traveler Pausanias.

By the end of the first century A.D., the Romans controlled a veritable industry of amber carving, with products ranging from small New Year's tokens of acorns and fruits to elaborate vessels for the tables of the wealthy. The principal center of manufacture was at Aquileia. Surviving objects, most of them from graves of Roman women, include toilet articles, rings, pendants, knife handles, mirror handles, small figures and figural groups, and small, highly carved containers and vases. The latter are often decorated with grape vines and putti in Dionysian themes. The best of these small vessels provide an idea of the famed, large luxury receptacles of amber that have all disappeared.

The Romans made much more extensive use of cloudy and opaque amber than did the Greeks or early Italic cultures, but it was carved as sparingly as the transparent pieces were. Still, the most highly prized type of amber was transparent and of the color that Pliny described as "like Falernian wine." Purportedly, Roman women often carried with them small lumps or stylized carvings of polished amber, simply for the enjoyment of its touch. By the third century A.D., the amber trade to Italy was drastically eclipsed due to the decline of the Roman Empire and warring Goths.

Overleaf, left: Mask of Dionysus. Roman, first century A.D. Height 4.5". Private collection

The piece is similar to one figured in D. E. Strong's Catalogue of the Carved Amber in the Department of Greek and Roman Antiquities *(British Museum); both have the backside with a large round plug that was probably fitted into the mouth of a vase. This piece, however, is much larger, more elaborate, and its additional bunches of grapes indicate it is Dionysus.*

Overleaf, right: Small vase. Roman, second century A.D. Height 3.7". Private collection

Carved from a single piece of amber that originally had some translucency and perhaps was dark amber to red in color, this vase now exhibits a slight ocher patina on the surface, due to nearly two thousand years of oxidation; otherwise, it is in superb condition. Represented in low relief are acanthus leaves and grapevines and tendrils that wrap along both sides, a signature Dionysian motif. A set of panpipes is carved in center bottom on one side. This piece is also similar to the one in Strong's British Museum catalogue, but that one has a pair of putti on each side and was probably suspended. This one probably stood free and contained an oil-based perfume; it gives an idea of the famed vessels of imperial Rome described in ancient sources.

MEDIEVAL AND RENAISSANCE AMBER

The history of Europe is reflected in Baltic amber, and perhaps no historical period involving amber in Europe imparts more mystique than that associated with the order of the Teutonic Knights, or *Deutschen Ritter*. The Teutonic Knights returned to Europe from the Crusades in 1211, and, in 1225, Conrad, Duke of Masovia, made an appeal to the knights for help in subduing the Balts along the Baltic coast. By 1283, the knights were absolute rulers of Prussia, and Baltic amber had become a lucrative commodity for trade with neighboring lands. It was extensively traded to the south, and, as early as 1302, shipments of amber were sent to the newly formed guild of amber, Paternostermachers (Makers of Lord's Prayer beads), in Bruges, where rosaries were crafted. In 1310, another amber guild was established, in Lübeck. By 1312, the knights had assumed a monopoly on the trade in Baltic amber. And trade burgeoned: in 1420, seventy guild masters and three hundred apprentices were listed in Bruges alone. Interestingly, a major stronghold of the Teutonic Knights at this time was in Marienburg. Today, the Teutonic castle in Marienburg (now Malbork, Poland) is a museum containing a collection of medieval amber objects.

The order strictly forbade the collecting of amber stranded on the beaches, except under the supervision of the Beach Master. Punishment was harsh and sometimes included hanging. Old woodcuts and engravings show licensed collectors harvesting the amber, with gallows in the background. (Especially notorious for instantly hanging anyone who pilfered amber was Anselmus of Lozenstein, judge of Samland. His spirit is said, by local legend, to wander the shore of Samland crying in penance: *Oh, um Gott, Bernstein frei, Bernstein frei* [Oh, by God, free amber, free amber]).

By the latter half of the fifteenth century, the influence of the Teutonic Knights had weakened, beginning in part with their defeat in 1410 at Tannenberg by the Poles and the Lithuanians. By 1466, the major Prussian cities were allied with Poland, and, in 1480, the king of Poland granted Danzig the right to its own Paternostermacher guild, which would eventually become the largest guild of amber workers.

In 1525, Albert of Brandenburg, the Hohenzollern ruler and grand master of the Teutonic Knights, secularized the order under the duchy of Prussia. His successor, Joachim II, converted to Lutheranism, and, in the next century, another Hohenzollern ruler, John Sigismund, converted to Calvinism, while his subjects remained Lutheran. Although Lutherans do not use rosaries, the

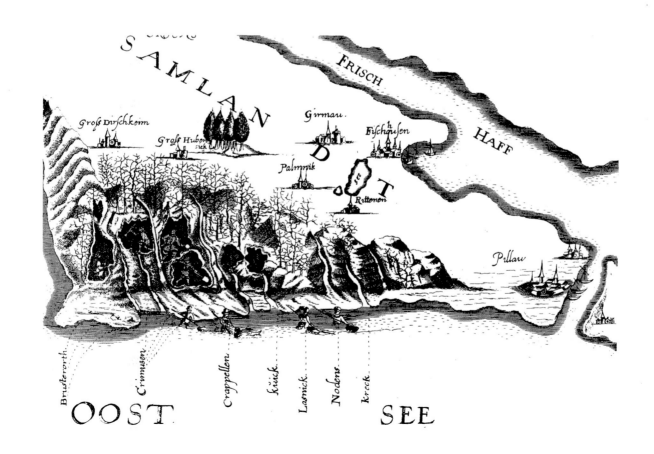

Etching from P. J. Hartmann, Succini Prussici physica et civilis historia, *1677*

Etching from P. Abraham, Etwas für Alle d. eine kurze Beschreibung, *1711*

Teutonic order struggled to maintain their control over amber by establishing their own Paternostermacher guild in Königsberg, which was to evolve later into the Royal Amber Works. The guild promoted secular items, such as game boards, goblets, cups, tankards, and caskets, and, especially in the fourteenth and fifteenth centuries, exported carvings of religious figures, cutlery with amber handles, crucifixes, and small bottles to French and Italian royalty. But the widespread decline in the use of rosaries after the Reformation ultimately led to the collapse of the order's amber guild.

From 1533 to 1642, the knights transferred the rights to the amber monopoly to the wealthy Danzig merchant family of Paul Koehn von Jaski. It was during the Jaski family's rule that the famous techniques of *Bernsteinstechen* (poking) and *Käscher* (catching) with long nets came into widespread use. Several old woodcuts and engravings, for example, show men in heavy leather cuirasses fighting the surf with their nets. Some drawings show fires suspended from poles onshore, supposedly used to thaw the gatherers' leather garments in very cold weather. A brisk trade to the Far East for Muhammadan rosaries kept the amber business

Rosary of amber and silver filigree beads with pendants and cross of silver filigree. German, seventeenth century. Length 25.8". Courtesy, Museum of Fine Arts, Boston. Bequest of William Arnold Buffum, 02.224

Fifty-five large, faceted amber beads and six beads of silver filigree are separated by sixty-three smaller amber beads that are irregular, some with a few facets but most with a rounded surface. All of the faceting is irregular. Only four decades are complete, and a fifth decade has twelve beads. Decades are separated by the silver filigree beads, and the second and fourth decades are interrupted by medallions in their centers. The large medallion suspended from the cross has central figures worked in gold showing the Coronation of the Virgin on one side and the Infant Christ with a Saint on the other. The medallion has an octagonal insert comprising small paintings of the Virgin on one side and Saint Veronica's Napkin (with the face of Christ) on the other, both under crystal covers. The smallest medallion contains side portraits, perhaps of the owner of the piece.

afloat, but barely. By 1642, amber was again Prussian property, and the strict rules of the Teutonic Knights were again enforced.

Paternoster beads, whose export was still the mainstay of the amber economy, have a rich history in themselves that deserves special attention. According to historical record, Paternoster beads, threaded in symbolic sets of numbers, were part of everyday life by the mid-thirteenth century. The *decade* of ten beads is the most commonly used, although sets of five and seven beads are also seen. Such sets, divided by marker beads (*gauds*), are used in repetitive prayers of penance and devotion. One early version of Paternoster beads was worn by men, who hung short sets of ten beads on a straight cord from the waist. Four fine sets made of amber are in the Musée Cluny, Paris.

The most pious individuals would make Paternosters simply by knotting a cord that hung from the waist; other versions for the humble were made of wood and bone. Beads of coral, crystal, and amber were particularly sought by royalty and the aristocracy. Guilds specializing in Paternosters of different substances—of glass and crystal from Venice, jasper from the upper Rhineland, coral from Paris and Barcelona, and amber from Lübeck and Bruges—formed throughout Europe. So prized were amber Paternosters that some Dominican and Augustinian friars in the thirteenth century spurned and even forbade their use, as a luxury and an excess.

Paternoster sets evolved into the present-day rosary in the second half of the fifteenth century. Valued Paternoster beads, including those of amber, were usually sent first to goldsmiths (such as in Gmünd), who fashioned the entire rosary and produced distinctive spacer beads of silver and gold filigree, and similar gauds in the form of medallions and crosses. Several fifteenth-century paintings of the Christ child depict the infant wearing rosaries of red beads, probably made of amber. In a 1510 edition of his poem "Le Triomphe des Dames," Olivier de la Marche wrote, "Moreover, my honoured lady must have pater-nosters of jet or coral, or for even finer ornament of fine amber."

Seventeenth–Nineteenth-Century European Amber

During the seventeenth and eighteenth centuries, northern Europe, particularly Prussia (now Germany and the eastern Baltic countries), was flooded with amber workshops and artisans. This is the period of the workshops headed by the great amber masters, who brought to the craft a sophistication never achieved before or since. Decorative objects, both secular and ecumenical, became grand: chests, inlaid cabinets, altars, inlaid game boards, groups of sculpted figures, lidded tankards, even several chandeliers, and, of course, the famous Russian amber room. With the exception of small sculptures and bowls carved from single pieces of amber, most objects were made by fitting pieces of amber together into a mosaic for an inlaid facade, or by joining with a central rod (in the case of intricately turned joints for candlesticks, bowl stems, and the like). None of this could have been achieved without vast supplies of amber, especially since pieces were individually selected for their color, size, and shape. It is no coincidence that the great workshops developed in towns along or near the southern coast of the Baltic Sea where amber was plentiful: Kassel, Lübeck, Danzig, and especially Königsberg.

The grand era of amber work also could not have developed without the sponsorship of a very elite aristocracy of the time. During the seventeenth and eighteenth centuries, absolutism triumphed in continental Europe, while England was becoming more democratic. It was the era of the czars of Russia, which began with the rule of Ivan the Terrible (1533–84) and flourished under Peter the Great (1682–1725) and Catherine the Great (1762–96). Prussia had become one of the most powerful European nations under the rule of Frederick I (1688–1713), Frederick William I (1713–40), and the "enlightened despot" Frederick the Great (1740–86). None of the European rulers, though, brought more opulence to nobility than did Louis XIV of France (1643–1715). His palace at Versailles took thirty thousand men more than twenty years to complete. Architecture, art, clothing, and decorative objects that inspired wonder and a disregard for the normal were the Zeitgeist among all European nobility, not just in France. Also, the ruling levels of the Catholic clergy were advisers and spiritual confidants to the nobility, and they adopted an equally rarified taste. The splendor and opulence of the age were captured in amber.

Amber was highly reworked in the Prussian workshops, so that various effects could be achieved. One of the most successful techniques was the use of inlays of ivory, meerschaum, and bony amber, which were inserted between small tiles and panels of amber or inlaid directly into the center of an amber piece. The creamy opacity of these materials beautifully contrasts with amber, especially deep-red transparent amber. In fact, bony amber was often used for the hands and faces of figures that were carved from a single piece of red amber. For many chests, small, intricately carved ivory reliefs were applied among the amber pieces. The scenes in these reliefs were commonly borrowed from well-known scenes depicted in Dutch engravings.

In another successful technique, similar to a glass-decorating technique called *verre églomisé*, a small panel of transparent amber was engraved in fine intaglio on the back side, then inserted into a mosaic. The intaglio was highlighted by a backing of bony amber, ivory, paint, or, more often, foil or mica. The effect takes perfect advantage of the natural properties of amber: a small window is produced, displaying a scene or phrase, creating focal points in what might otherwise be a monotonous mosaic of amber pieces. The greatest use of this technique was in game boards, which could have as many as one hundred such windows. French inscriptions made this way indicate origins from Danzig workshops.

The other major innovation was *encrustation,* in which squared wafers of amber were glued to a wooden frame. This technique began in the mid-seventeenth century in Danzig, and it soon spread for use in constructing elaborate chests, cabinets, and altars, some of imposing size. Prior to this, the size of chests was constrained by the size of the amber tiles that were dovetailed to form the actual wall of the chest. With any of the chests, reliquaries, or altars, the facades were always composed of a mosaic of amber pieces varying greatly in color, from cloudy to fatty amber, and transparent yellow to red. By the early eighteenth century, elaborate turning of amber pieces was also developed to create a delicate and intricate structure. By boring a hole through the center of the smaller turned pieces and inserting a metal rod, an artisan could join all the elements to produce a strong shaft but with delicate proportions and translucency.

The most productive centers of amber work were Danzig, where the amber guild was founded in 1477 (it still exists), and the Prussian cities of Kassel and Königsberg, the latter of which established an amber guild in 1641 (and which lasted until 1811). Establishing the provenance of seventeenth- and eighteenth-century European amber pieces is greatly complicated by a rarity of signed and dated works, but, despite the proximity of Danzig to Königsberg (only 75 miles), the products were quite different. Catholic Danzig largely produced small house altars, shrines, and reliquaries, saints and other religious figures, and pieces for the celebration of the Mass, such as cruets and chalices. Protestant Prussia produced mainly secular items: cabinets, small chests, figures of mythical Greek and Roman heroes, game boards, lidded tankards (*Deckelhumpen*), stemmed bowls, and snuffboxes. One can imagine the table of an aristocratic Prussian

Arched wood panel with inserts of transparent amber medallions and buttons, comprising a total of 521 amber pieces of various colors and transparency. Italian, nineteenth century (panel), some pieces earlier. 26.5 x 17". Courtesy, Museum of Fine Arts, Boston. Bequest of William Arnold Buffum, 02.222

This devotional panel was probably used in a household or small church. The central figures at the bottom, representing Joseph (left) and Mary (right) adoring the Christ Child held in the arms of an angel, are carved from an unusual opaque and mottled brown amber, possibly a composite, which is now friable and broken. "Windows" of copal panels (extensively crazed) lie behind these figures and those in the smaller arches to the sides. In the medallion just above the three figures is a man being visited by an angel. Above that one, in a niche, is presumably the Christ Child, throwing off a cloak. Above the figure in the niche is an octagonal medallion with a relief of two shepherds. Two arched niches on each side have (left) a female figure resembling Liberty and (right) a woman with a staff leading a child by the hand. Above these figures in the pair of niches is a small medallion with a side portrait. Two octagonal medallions below the side niches have scenes inscribed on the back of the amber, of towns near bays. Beneath these octagonal medallions are oval ones with figures of men carved in relief: the left one, of a "musketeer," the right one of a soldier. This work is very similar to a square panel, also in Boston; both are uncommon pieces.

The Virgin. German, eighteenth
century. Height 3.9". Courtesy,
Museum of Fine Arts, Boston. Bequest
of William Arnold Buffum, 02.239

In this beautifully carved figure, the
Virgin, in flowing robes, with hands
folded over her breasts, looks upward
as if to receive a divine message. As
are all of the figures reproduced from
the Buffum Collection, she was carved
from a single piece of clear Baltic
amber fastened to an ivory base with
a thin ivory peg fitted into the bottom
of the figure.

Figure of a woman. German, eighteenth century. Height of figure 3.5". Courtesy, Museum of Fine Arts, Boston. Bequest of William Arnold Buffum, 02.241

The woman carries a box in her left hand.

Figure of a saint. German, eighteenth century. Height of figure with base 3.3". Courtesy, Museum of Fine Arts, Boston. Bequest of William Arnold Buffum, 02.243

The bearded man, holding a staff in his right hand and the infant Christ cradled in his left arm, perhaps represents Saint Christopher. He is wearing boots. The base was made from a separate piece of reddish amber.

Altar. Danzig, c. 1775–1800. Height 47". Victoria and Albert Museum, London

This imposing altar on four ivory feet has scrollwork along its sides in a distinctive Danzig style. The front of the altar depicts in great detail scenes from the New Testament, some adopted from Dutch engravers. At the base are twelve figures of the apostles. The resurrected Christ is at the top. In the back of the altar is a clock calendar, made of delicate ivory carvings, with some strips of amber added. The back scenes depict the zodiac, Adam and Eve, and additional New Testament scenes. The only piece comparable to this is one by Königsberg amber master Georg Schreiber, in the Museo degli Argenti, Florence. Another, in Malbork Castle, is less elaborate.

Opposite: Crucifix. Poland, late seventeenth century. Height 32". Russian State Museum, Saint Petersburg

One of the few crucifixes made of amber that is largely intact, this piece is close in style to the work of Prussian master Friedrich Schmidt. Like most crucifixes of the period, it is very intricate and fragile, with delicately carved scenes depicting Christian feasts. The bottom tier has sliding panels that open to reveal the Holy Sepulcher within. The piece belongs to the collection of the Russian State Museum but is housed in Tsarskoye Selo, near Saint Petersburg.

Nine chess pieces. German, seventeenth century

Details from chessboard on page 142

Above: Amber playing square:
L'un et l'autre la crou (*"The one and the other cross"*)

Below: Amber playing square:
À la guerre et à la paix (*"To war and peace"*)

Pair of cruets (alpha and omega) for the Mass. German, eighteenth century. Heights 5.1" each. Courtesy, Museum of Fine Arts, Boston. Bequest of William Arnold Buffum. 02.226 (alpha), 02.227 (omega)

Each cruet is made up of six pieces of deep-red transparent amber: the handle; the Greek letter; the neck with spout; the central, ribbed section; the piece just below, with parallel engraved lines and the carved ball; and the base with radiating carved lines. Some crazing has occurred, and the shape of the handle of the omega cruet, which was badly damaged and restored, has become distorted. The omega letter itself was fashioned from a new (opaque brown) piece of amber, perhaps after acquisition of the Buffum Collection in 1901. Despite the damage, these delicate pieces are in remarkably good condition and are among the finest objects of their kind made from amber.

Chest. Danzig, from the workshop of Gottfried Turau, dated 1705. Length 16.2". Ekaterininsky Palace Museum, Saint Petersburg, 60-VI

A beautifully symmetrical and intact lidded chest made of a wooden frame with amber mosaic carved in relief. Acanthus leaves and transparent oval medallions (with engraved designs) decorate the front. The carved group on top represents Venus and Cupid. The inside is lined with velvet. During restoration in 1983, the artist's signature was found on the wooden base: GOTTFRIED TURAU, INVENTOR ET FECIT ANNO 1705 12 JULIUS.

Crouching male lion. German, eighteenth century. Height 1.3". Courtesy, Museum of Fine Arts, Boston. Bequest of William Arnold Buffum, 02.204

Between his paws, the lion holds a globe carved from a piece of transparent yellow amber. The small figure probably once rested on top of a lidded container; a very similar one exists in Berlin.

Below: Snuffbox. German, eighteenth century. Diameter 1.97". Courtesy, Museum of Fine Arts, Boston. Bequest of William Arnold Buffum, 02.202

Both bowl and lid were probably fashioned from a single piece of deep-red, transparent amber. Carved in high relief on the lid is a winged cupid sitting on rocks with a pastoral scene in the background. Along the bottom edge is a fleur-de-lis relief. Inside is a smooth bowl of brass.

Above: Case for needles or toothpicks. German, first half of eighteenth century. Height 3.5". Courtesy, Museum of Fine Arts, Boston. Bequest of William Arnold Buffum, 02.201

The base was carved from a single piece of amber, the top two-thirds opaque; the cap was made from a separate piece of bastard amber; the collars are of copper and brass.

Middle: Wedding knife and fork. German, eighteenth century. Length 7.3" each. Courtesy, Museum of Fine Arts, Boston. Bequest of William Arnold Buffum, 02.210 and 02.212, respectively

The handles are made of clear red amber, with a steel shaft through each, the knife being a carved head of a man, the fork a head of a woman.

Below: Small box with four counters. German, nineteenth century. Diameter 1.4". Courtesy, Museum of Fine Arts, Boston. Bequest of William Arnold Buffum, 02.209

The lid, bowl, and counters are made of transparent, slightly cloudy yellow amber. Carved out of the centers of the wafer-thin counters are a heart, diamond, spade, and club. Each counter has a wreath of flowers and buds carved on one side. The cover displays a pattern of rays, at its center a button of slightly clearer amber containing a caddisfly inclusion; the bottom of the base incorporates a similar pattern of rays.

obsessed with amber, serving a dinner to impress guests: The plates might have had inlays of amber; the handles of the knives and forks might have been carved from amber; passed around the table, perhaps, was a small bowl, made from amber, containing sweets; one might have drunk from an elaborate tankard crafted from amber, similar in color to the beer it contained; and light from the table could have come from candles mounted in candlesticks of amber. After dinner, toothpicks might come from an amber needle case, snuff from an amber snuffbox, and a round of chess played with pieces carved from amber and a board of intricately inlaid amber.

Perhaps the most famous amber master from the Danzig workshops was Christoph Maucher (1642–after 1701), whose work embodies the Danzig style but who chose not to become a member of the Danzig amber guild. He received many prestigious commissions from the Brandenburg court, and this, in turn, inspired jealousy among his amber colleagues. He is particularly well known for his subjects and style of carving amber. Several of his figures, whether completely carved or in deep relief, depict, for instance, the Judgment of Paris. Although they are part of classical scenes, the figures have appealing, earthy proportions and gestures. Such works were generally carved from a single piece of opaque amber, which originally must have been at least eight inches in diameter, and they were used for the tops of cabinets encrusted with amber. Maucher had an associate, Nicholas Turow, who was commissioned in 1677 by the Elector of Brandenburg to craft a magnificent throne of inlaid amber for presentation to Emperor Leopold I of Austria. Only fragments of it remain in Vienna.

Hallmarks of Danzig amber work are the small altars and tabernacles, built in tiers with arabesque scrollwork on the sides, similar to tomb monuments in the city. Michel Redlin, another Danzig amber master, was especially renowned for his elaborate chests and altars. He is also one of the very few masters whose drawn plans for several pieces still exist.

The history of amber work in Königsberg and nearby Prussian towns is very distinguished. Among the earliest masters was Stenzel Schmidt, who worked in the late sixteenth century, and of whom little is known. Perhaps the most celebrated of any amber artisan was Georg Schreiber (active c. 1615–43 in Königsberg), or, as he is known from the few works he signed, "Georgius Scriba." His pieces represent the finest standards in amber work, from the architectural designs of chests and altars to the inlays, intaglios, and reliefs carved into the amber and ivory plaques of various decorative objects. A magnificent lidded tankard in Darmstadt, Germany, is one of his rare signed pieces; a signed game board (dated 1616) with a complete set of amber chess pieces sold in 1990 at Sotheby's of London for £330,000. Among his masterpieces, and unusual for his genre, is a signed altar with an imposing crucifix, dated 1619, in the Museo degli Argenti, Florence. The elaborate composition and intricate delicacy of craftsmanship are probably the finest of any amber decorative piece. Like most of the superb items in the Museo degli Argenti, this masterpiece is in fine condition.

Pair of earrings. Sicilian, nineteenth century. Length 2.9" each. Courtesy, Museum of Fine Arts, Boston. Bequest of William Arnold Buffum, 02.217, 02.218

Now crazed, the transparent yellow amber used in these amber drops mounted in mercury-amalgam gilded silver may be from Sicily or the Baltic region.

Opposite, left: Candlestick. German, seventeenth century. Height 10.8". Courtesy, Museum of Fine Arts, Boston. Bequest of William Arnold Buffum, 02.228

The top portion, a dish with a central spike, is brass. The stem is made of six turned pieces of reddish transparent amber, each with a hole bored through its length into which a brass shaft has been inserted; the pieces are tightened by screwing a small bolt at the end of the shaft, beneath the base. The square base is not a veneer of amber but is made entirely of twenty-six panels of transparent amber supported by a thin brass frame at the very bottom. The base has four corner pillars; between each is a panel with a small arched window. The windows originally displayed a gold-leaf design behind them, which suggests that they were made of transparent amber. They were replaced with the opaque window pieces that presently are in place.

Opposite, right: Perfume bottle. German, nineteenth century. Height 5.7". Courtesy, Museum of Fine Arts, Boston. Bequest of William Arnold Buffum, 02.229

A highly polished object made from five turned pieces of transparent amber, its base and neck pieces are deep red, and the body, cap, and top button are yellow amber. The interior is also highly polished.

Another Königsberg master was Jacob Heise (active 1654–63), who is particularly well known for his bowls, cups, and lidded tankards, one of which is at The Metropolitan Museum of Art, New York. He also perfected the crafting of stemmed bowls. The bowls and tankards are made in a similar technique. They are composed of six to ten panels of clear amber that are fused along their edges to form the barrel of the bowl or tankard (for the tankards, the panels are virtually flat; they are very curved for the bowls). The stems of the bowls are made of several pieces of turned and carved amber secured with a rod running through the center. The top and bottom of the tankards are made of clear amber as well, sometimes from a single piece. For both the tankards and the bowls, laced gold trim adorns the edges, and the surfaces of the panels are carved in intricate shallow relief. Some items have ivory or bony amber inlays.

Jacob Dobberman (active 1716–45), an amber master from Kassel and court artisan for Karl I of Hesse, is renowned for his chandelier of many turned pieces of amber. Another, much more intricate chandelier, with numerous faceted pendants of amber, by Lorenz Spengler, is in Rosenborg in Copenhagen. Spengler (1720–1807) was master to the Danish royal court. Like many amber masters, he worked in both ivory and amber. In fact, he carved two identical pairs of putti, now in Rosenborg, one of ivory and the other of amber. A signature design of his work is a crown on a pillow, which adorned the chandelier and larger objects such as chests.

Major collections of European ambers are at Rosenborg, Copenhagen; Museo degli Argenti, Florence; the Victoria and Albert Museum, London; Malbork Castle, Poland; Residenzmuseum, Munich; Kunsthistorisches Museum, Vienna; Skokloster, near Stockholm; Tsarskoye Selo, Russia; and the Museum of Fine Arts, Boston. The most impressively large and delicate pieces are in Munich, Vienna, and, especially, Florence. The Boston collection, which is illustrated here, is unexpectedly rich for a North American collection. It was bequeathed to the Museum of Fine Arts by Arnold Buffum in 1901 and consists of sixty-two pieces and sets of mostly Baltic amber, some purportedly of Sicilian and Romanian amber. Buffum was an archaeologist and amber connoisseur who is also known for his small, popular book that romanticized the European lore and use of amber, *The Tears of the Heliades, or Amber as a Gem* (1897). In the mid- to late nineteenth century, he traveled extensively around Italy and other parts of Europe collecting the objects. Pieces from the Buffum Collection have a great mixture of provenance and dates, much of which can only be inferred, but this adds to the collection's charm: it is very representative of the spectrum of European artistry in amber.

Pendant. English, nineteenth century. Height 1.8". The Natural History Museum, London

Amber jewelry with insect inclusions was rare in Victorian England, where Baltic amber jewelry was in vogue. This beautiful, clear piece of Baltic amber surrounded by diamonds contains a cricket and a spider.

Opposite: Necklace, brooch, and
earrings. Italian, c. 1860–70. Courtesy,
Museum of Fine Arts, Boston. Bequest
of William Arnold Buffum, 02.91
(necklace), 02.92 (brooch), 02.93/94
(earrings)

These cabochons of highly polished
Sicilian amber set in gold pendants
reflect an embellishment of a classical
design. The set purportedly was made
by the Castellani firm of Rome, or in
a manner very similar to Castellani's,
specifically for Arnold Buffum.
Castellani jewelry in the nineteenth
century reflected a revivalist movement
inspired by Etruscan designs. One
hallmark of this design is the
distinctive granulation in the gold.

Game board with crown on a pillow.
Probably Copenhagen, workshop of
Lorenz Spengler, mid-eighteenth
century. Length 3.2". Ekaterininsky
Palace Museum, Saint Petersburg, 7-VI

With the exception of the yellow cross
and orb on top (which is a modern
addition; the original was lost during
the evacuation of 1941), the crown is
carved from a single piece of opaque
amber, the pillow from a separate
piece. Of a subject and style distinctive
to Spengler, the piece is also significant
as the last acquisition made for the
Tsarskoye Selo amber collection,
acquired in 1912 by Czar Nicholas II.

THE AMBER ROOM

*T*he pinnacle of amber creations is the famed eighteenth-century amber room from Russia. Inspiration for it probably derived from the chests (caskets) and house altars of inlaid amber, sometimes of elaborate architectural design, that were in vogue in Königsberg and Danzig in the seventeenth and eighteenth centuries. Although the room resided in Russia for more than two hundred years, it was actually a Prussian creation. The story of its origin, remodeling, and eventual loss is an intriguing one.

In 1701, King Frederick I of Prussia commissioned for the main palace in Berlin a banquet room with panels of amber. The original commission went to the Danish amber "cutter" Gottfried Wolffram, but in 1707, the Danzig amber cutters Ernst Schact and Gottfried Turau were hired to replace Wolffram, who was proving too expensive. Panels (with a base of oak) were said to have been made of 100,000 pieces of amber laid into mosaics of floral design, royal heralds, and profiles. Each panel also had the Prussian coat of arms (a crowned eagle in profile). Soon after its completion in 1712, the room was seen and admired by Czar Peter I (Peter the Great).

In 1716, Frederick William I of Prussia (son of Frederick I) signed the Russo-Prussian Alliance with Peter I, against Karl XII of Sweden. As the Russian poet Aleksandr Pushkin wrote: "The haughty Swede here we'll curb and hold at bay. And here, to gall him, found a city [Saint Petersburg]." To commemorate the alliance, the amber room was presented to Russia in 1717. (It was rumored that Frederick William I had a distaste for such opulence, anyway, and he knew that Peter I had admired the room.)

The first home for the amber room in Russia was in the Old Winter Palace in Saint Petersburg, where the panels lined a study. These panels were moved in 1755 to the Ekaterininsky Palace in Tsarskoye Selo (Czar's Village). Elizabeth, daughter of Peter I, had the architect Varfolomei Rastrelli design a room especially for the amber panels. Originally the room contained twelve wall panels and ten "pedestal fields." Four of the pedestal fields were inscribed with Frederick I's initials, four with the Prussian eagle, and two with emblems representing armed peace. The new room was much larger than the one in which the panels had first been installed. For example, it had a thirty-foot ceiling, instead of the original ceiling of sixteen feet. This must have been when panels emblazoned with the Romanov crest were added. Rastrelli framed each panel in an elaborately gilded cartouche and inserted twenty-four mirrors between

The amber room, photographed in the 1930s,
before the room was disassembled by the
Nazis. Where the panels are today is still
unknown, although the amber decorative
objects were saved by the Russians.

Above and opposite: Details of replicated panels from the amber room. Ekaterininsky Palace, Saint Petersburg

them. A ceiling mural by Francesco Salvatore Fontebasso and a wood parquet floor were also added.

Until at least 1763, five Königsberg amber masters continued to work on the amber room. The group was headed by Friedrich Roggenbuch and included Johann Roggenbuch, Clemens and Heinrich Wilhelm Frick, and Johann Welpendorf. Like many aristocrats of the day, Peter I owned a *Kunstkammer,* or cabinet of curiosities and wonders, which featured amber artifacts. Those artifacts graced the amber room along with ones created by these artisans. By 1765, the number of objets d'art made of amber had grown to seventy, including small chests, candlesticks, snuffboxes, saucers, knives and forks, crucifixes, and tabernacles. (One of these original pieces—a splendid late seventeenth-century crucifix by Polish amber master Friedrich Schmidt, depicting feasts in intricate relief—is in the collections of the Hermitage.) In 1780, a small corner table, made of encrusted amber with a large cabochon of clear amber just above the leg, was added. The last amber accession was in 1913, just prior to the revolution. It was an amber crown on a pillow, symbol of the eighteenth-century master Lorenz Spengler. The piece, which had formerly adorned the lid of a chest, was purchased by Czar Nicholas II.

The amber panels required restoration over the years, presumably to replace pieces that had fallen from the mosaics or had become heavily oxidized. Restorations were done in 1810, 1830, and 1911. After the revolution, tours of the room were given to people who were required to wear felt slippers. One of the last persons to have seen the intact amber room described it as "stepping into a fairy tale."

When the Nazis invaded Leningrad (formerly Saint Petersburg) in 1941, the Russians had already removed the small objects from the amber room to Novosibirsk for safekeeping and quickly papered over the amber panels in a desperate attempt to hide them. According to Alfred Rohde's 1942 article on the room, two Nazi officers on the Russian line, Count zu Solms-Laubach and Captain Poensgen, were art historians in their civilian lives. They realized the value of the room and knew that, in order to save them from bombing, the panels would have to be removed. It took six men thirty-six hours to dismantle the room. The panels were crated to Kaliningrad (old Königsberg) by train and installed in the main castle along with vast collections of other art. Rohde reported that the room was "returned to its true home [Germany], the real place of origination and sole place of origination of the amber [Prussia]. The amber room of Frederick I is next to the hall of Louis Corinth." It did not stay there long.

Presumably, in 1945, as Kaliningrad was being bombed by the Western Allies, the panels were again taken down and hidden. An S.S. commander, Erich Koch, was apparently the only person who knew of their existence, since he had been in charge of the panels, but he died in 1986 without revealing their location. International intrigue surrounds the possible existence and location(s) of the panels. Many believe they were destroyed either by fire or bombing. One imaginative explanation, published on September 24, 1994, in the Erfurt-based newspaper *Thüringer Allgemeine,* quoted former German officer Gert Sailer and a Russian art historian who claimed that the panels were smashed in the basement of the Königsberg palace by drunken Russian soldiers celebrating their reinvasion of the city. Why the soldiers would not have recognized the Baroque masterpiece is unexplained. Various hypotheses also exist as to where the panels may still be residing: in bunkers beneath Lubbenaw or Weimar, Germany; in a lost subterranean ice room of the brewery that Koch frequented; or in mine shafts in Ohrdruf.

Russian artisans began replication of the amber panels in 1979, the plan being that reproductions would be exact down to each piece of amber. The replicas being made are based on large black-and-white photographs taken before the Nazi invasion, which show details of inlaid mosaic. The contrast between the bony, bastard, and clear amber pieces makes outlines of each piece discernible. Seventy pieces of amber from the panels that had been left behind in the Germans' haste provided information to restorers as to how the original amber was prepared by "clarification" in hot oil and sometimes even dyed. Each amber piece is carefully cut to shape and thickness, polished, and fit into the mosaic by master amber crafters. Wood carvers and gilders are working on the elaborate cornice work. Despite Boris Yeltsin's announcement in November 1991 that the amber room is hidden in Germany (no evidence was given), the restoration continues. The work is impeded by the cost of so much amber, which still comes from the famous Palmnicken mines.

Right: Table chest of drawers. Danzig, first half of eighteenth century. Length 21.7". Ekaterininsky Palace, Saint Petersburg, 61-VI

Composed of a wooden frame with an inlaid mosaic of amber, the chest features seven drawers in two rows, the large central one with two heraldic lions and engraved scenes of Danzig. On the lid is a composition made in the pietre dure *technique, showing a bird perched on a branch. Received in Tsarskoye Selo in 1765, it is the only piece of furniture from the room to have survived to the present time.*

Below: Shaving basin with soap dish and brush. Tsarskoye Selo amber workshop, 1767. Length of basin 11.4"; height of dish 3.9"; length of brush handle 4.3". Ekaterininsky Palace, Saint Petersburg, 15-VI (basin), 127-VI (dish), 43-VI (brush)

The deep oval basin was made from a supporting frame of pounded sheet metal to which panels of amber were fixed. A relief is inscribed in the bottom oval panel. The pineapple-shaped soap dish features a scaled relief and a detachable lid. The handle of the shaving brush was turned from a single piece of amber. These objects were part of the so-called Knight's Set, which originally included a case for ointments, a snuffbox, two studs, a flask, a jar, a toothbrush, and a clothes brush, all made from amber.

Opposite: Dishes with figures in relief. Kassel, workshop of Johann Christoph Labhard, first half of eighteenth century. Lengths 3.9–5.1". Ekaterininsky Palace, Saint Petersburg, 8-VI, 9-VI, 12-VI

A popular genre of the time, these dishes often depict kissing couples or female nudes with Satyrs. One, of opaque yellow amber, is an allegorical figure of Mercy with two putti.

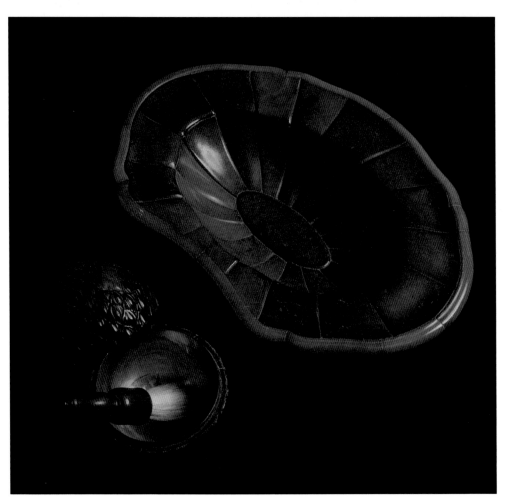

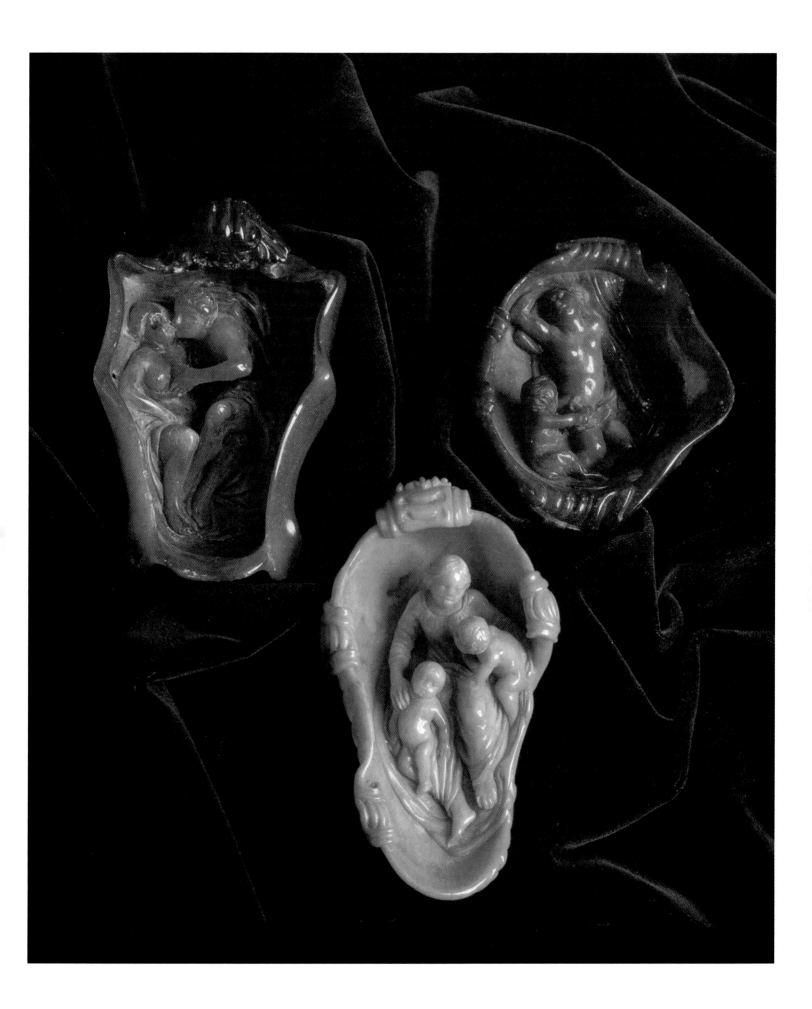

ASIAN AMBERS

Besides Europe, only China and, to a lesser extent, Japan possess a rich history of the artistic use of amber. While European folklore on the origins and properties of Baltic amber was growing, an independent lore about amber developed in China. The first mention of amber in Chinese literature is from the *Ch'ien Han shu (Annals of the Former Han Dynasty)*, by Pan Ku, in A.D. 85. There is a much later mention of "Hsi" amber, in the fifth century A.D., but that actually refers to jet. The Chinese, like the Europeans, put forth accurate as well as embellished explanations for the origin of amber. The Chinese were aware of amber's botanical origin at least as early as the fifth century A.D. T'ao Hung-ching dismissed the myth that amber is formed from burned bee nests: "There is also amber, in the midst of which there is a single bee, in shape and color like a living one. The statement of the *Po wu chih*, that the burning of bees' nest effects its make, is, I fear, not true. It may happen that bees are moistened by the fir resin, and thus, as it falls down to the ground, are completely entrapped." In the second half of the eighth century A.D., Li Hsün wrote: "Amber is a secretion in the wood of the sea-fir. At first it is like the juice of the peach tree; later it coagulates and assumes form." And in the middle of the tenth century A.D., Han Pao-shêng wrote in his *Shu pên ts'ao,* the medical text of Szechwan Province: "The resin of the liquidambar-tree penetrates the earth, changes during a thousand years, and thus becomes amber."

Oddly, much later, the origins of amber surface in mythology. As outlined in the *Pên ts'ao kang mu* of Li Shih-chên, a sixteenth-century work on natural history: "When a tiger dies, its soul penetrates the earth, and is a stone. This object resembles amber, and is called *hu p'o* [tiger's soul]. The ordinary character is combined with the radical *yü* [jewel or, literally, 'jade'], since it belongs to the class of jewels."

Most of what we know about the early Chinese trade in amber comes from Berthold Laufer, a famous German-born sinologist. Virtually all of the Chinese objets d'art in amber come from Beijing, which is not to say this is the original provenance of the pieces. The origin of the amber, though, is clear; it came from Burma, at least prior to the nineteenth century. China does have some large natural deposits of amber in Fushun, but these appear not to have been exploited hundreds of years ago, let alone used in decorative objects. The burmite from northern Burma was probably brought by Chinese traders to Yunnan Province as early as the first century A.D.

Opposite, above: Skull carved from subfossilized wood impregnated with resin. Possibly from Tibet, eighteenth century. Height 1.7". American Museum of Natural History (Anthropology). Drummond Collection, 70.3.2700

The skull is mostly black; the lighter areas are tan and resemble extremely dense wood, with growth rings showing.

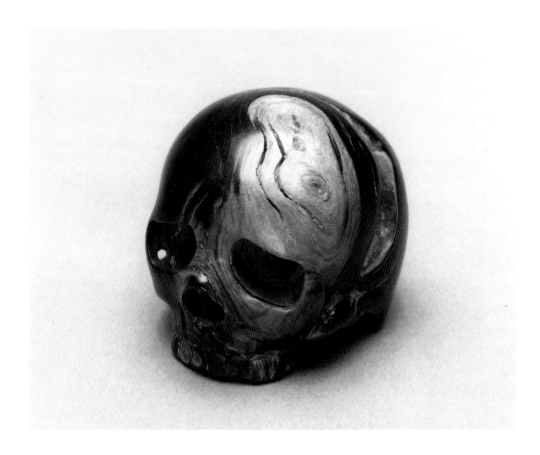

Below: Carving of a lotus root, flower, and leaf, with two goldfish. China, eighteenth–nineteenth century. Height 3". American Museum of Natural History (Anthropology). Drummond Collection, 70.3.2582

Pairs of fish often denote a joyous union, such as a married couple. A small goldfish is in the center of this piece; the tail of the larger one is intricately carved. The reverse side has much less relief but does feature inscribed details of scales and leaf veins. Transparent in places, the sculpture is mostly opaque.

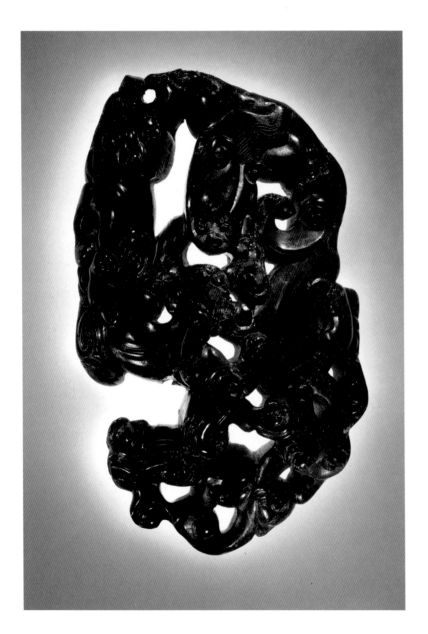

Opposite, left: Two dragons. China, eighteenth century or earlier. Height 2.7". American Museum of Natural History (Anthropology). Drummond Collection, 70.3.2702

The most important creature of Chinese mythology, the dragon is considered beneficent and omnipotent, a symbol of vigilance and imperial power. Here, two very stylized and intricately carved dragons, their bodies intertwined, face each other in the center of what probably was used originally as a pendant. The amber, which had been cracked in the center, is distinctively dark and opaque.

Opposite, right: Calligraphic character. China, seventeenth–eighteenth century. Height 2.3". American Museum of Natural History (Anthropology). Drummond Collection, 70.3.2580

The character is a blessing, a popular subject for jade carvings. The surface between its edges is slightly depressed, due to the high polish of the edges. The walls of the holes, which were drilled and cut, were not polished.

Left: Chinese "Madonna." China, eighteenth century. Height 4.5". American Museum of Natural History (Anthropology). Drummond Collection, 70.3.2707

Carved from a single piece of opaque amber, mounted on an ebony stand, the figure (Kwan-yin, goddess of mercy?) originally must have held a staff, which is now gone. Her hair is rendered in very fine parallel lines. A small purse hangs from the left side of her waist. She perhaps represents a goddess to whom a woman who wished for a child might pray.

Various scattered references attest to the value of amber used in cherished objects, from China and elsewhere. One of the princes of the state of Wu, Sun Ch'üan (A.D. 181–252) is said to have owned a scepter of amber, which he accidentally broke in his delight over the portrait of a beautiful woman. The K'ambra women of Tibet reportedly wore head ornaments with a large oval piece of amber in them, and the Dalai Lama once made a gift of amber to a Chinese

Opposite: Carp leaping from the water. China, early eighteenth century. Height without base 2.7". American Museum of Natural History (Anthropology). Drummond Collection, 70.3.26.98

Carved from a single piece of deep-red, transparent amber, sitting in an ivory base, the fish is recognizable as a carp by the barbel near the mouth. Carp are associated with martial arts and with perseverance because of their ability to struggle against the current of a river; this one leaps from waves. Carving appears on both sides and includes inscribed scales and waves. The tail is more complete on one side. The hole near the mouth indicates that the piece was also worn as a pendant.

Below: Jui-Ruyi. *China, eighteenth century. Length 14". American Museum of Natural History (Anthropology). Drummond Collection, 70.3.2511*

A wish-granting scepter, this decorative piece was used to promote good wishes at special occasions like birthdays and weddings. It apparently was made from a single piece of opaque, mottled reddish-brown amber, although the stem was broken and repaired in four places. Along the stem are intertwined branches. Bracket fungi (Lingzhi), which signify longevity or immortality, are carved at the base and near the neck, with flowers and buds between. The neck is branched into two limbs, just beneath the crowning "bouquet" of bracket fungus, on which sit a smaller bracket fungus, a magnolia flower, bud, and leaves. Partly protruding from between the two branches is a bat, with the head and one wing showing. Many ruyi come from the Qing dynasty.

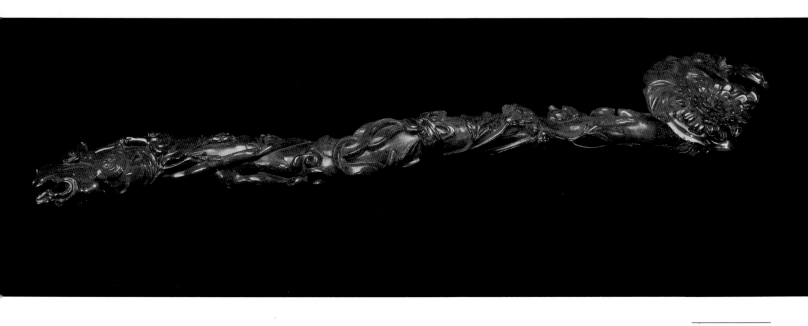

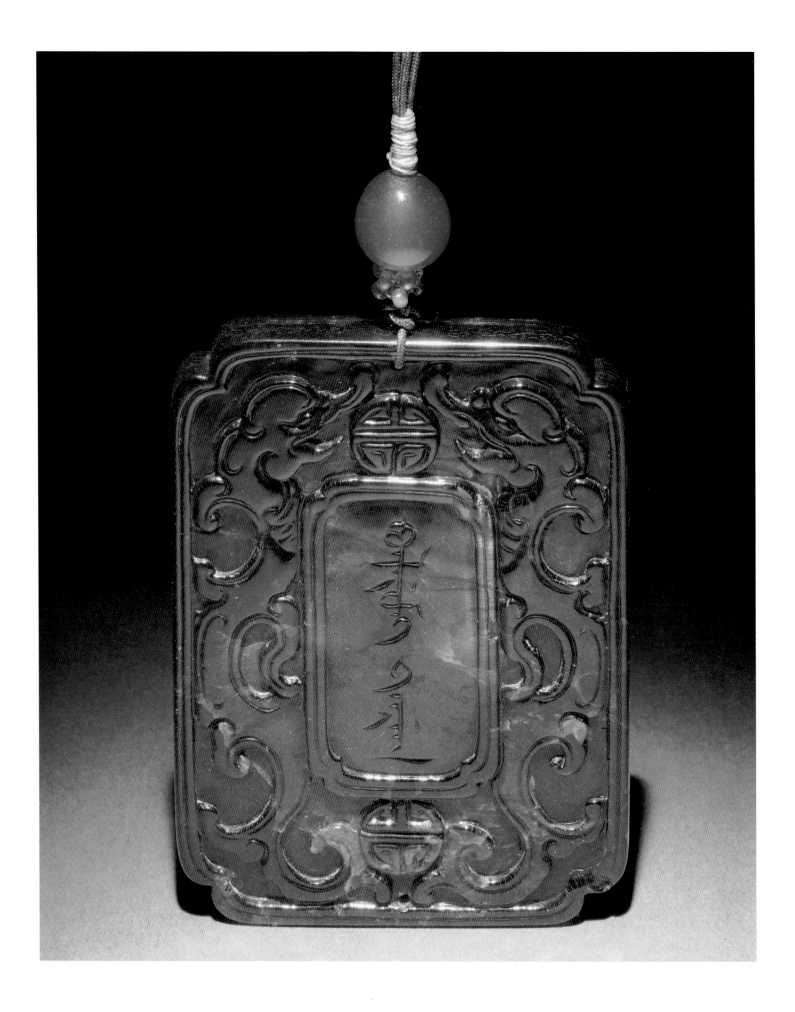

Opposite: Pendant. China, eighteenth century. Height 2.5". American Museum of Natural History (Anthropology). Drummond Collection, 70.3.2597

The pendant is largely of transparent red amber with some inclusions of debris. Surrounding the central rectangle, on both sides, are a pair of facing stylized dragons. Inscriptions also appear on both sides, one in Manchu script (Qing dynasty), the other in Chinese. Suspended from silk cords with a bead and cluster of seed pearls, it was worn by court ladies as a symbol of abstinence.

Dragon pendant. China, nineteenth century. Height of amber 2.1". American Museum of Natural History (Anthropology). Drummond Collection, 70.3.2628

Made from transparent amber, this coiled dragon features forked ears and tail. A bead of tourmaline and two clusters of seed pearls are attached to the silk cord.

emperor. According to the *Old History of the Tang Dynasty,* the king of Nepal adorned himself with pearls, mother-of-pearl, coral, and amber. All of these references are almost certainly to burmite. One reference, however, probably concerns a gift of Japanese Kuji amber, the size of a "peck," given by Japanese emperor Toyoshi-karu no-oji to Chinese emperor Kao Tsung of the Tang dynasty (around A.D. 650).

The early trade routes of burmite indicate that Chinese traders transported the material to Yunnan Province from northern Burma via Mogung and Myitkyia. About 1633, Portuguese Jesuit priest Alvarez Semedo wrote of his travels in southern China: "Yunnan is a great countrie, but hath little merchandise . . . whereof they make the beads for chapplets, which in Portugall they call Alambras; and in Castile, Ambares; and are like amber, they are counted good against catarre; it is digged out of mines, and sometimes in great pieces: it is redder than our amber [i.e., Baltic], but not so clean. . . . They are excellent in works of ivory, ebony, and amber." This description is obviously of burmite, at that time still more easily obtained than Baltic amber. By the nineteenth century, European trade with China made Baltic amber more available in China than burmite, although the latter material was imported into China as late as the 1930s.

Among the collections of Chinese amber objects in Western museums are those at the Victoria and Albert Museum, London; the Packard Collection and other pieces at the Museum of Fine Arts, Boston; and, probably the largest in the world, the Drummond Collection at the American Museum of Natural History, New York. The Victoria and Albert collection consists of nine pieces of eighteenth- and nineteenth-century origin, including a necklace, small vases, several bottles, saucers, and figurative carvings. The Museum of Fine Arts collection includes an eighteenth-century Qing dynasty carving of peaches (most likely made of copal), Qing pendants, and an eighteenth-century opaque yellow piece depicting birds, flowers, and a dragon. The finest pieces in that collection are the Sakyamuni Buddha, made of opaque, dark red-brown amber from the late Ming dynasty (seventeenth century), and an exquisite octagonal cup from the Qing dynasty. The cup is carved from opaque yellowish amber in an archaic bronze form with two handles, each in the shape of a dragon. The Buddha and cup are made of amber quite atypical of burmite and are probably crafted from Baltic amber.

The collection at the American Museum of Natural History, bequeathed in November 1933 by Isaac Drummond, consists of 1,893 decorative objects in jade, ivory, and amber. Drummond had assembled part of the collection himself and inherited other portions from his father, James F. Drummond. The collection is a prime resource for comparing how amber was carved in the manner of jade in both subject and style. It also demonstrates that the variety of Chinese objets d'art in amber is nearly comparable to that of European objects. The amber pieces alone number several hundred pendants, bottles (a favorite of Drummond's), figures, bowls, and small house objects such as an inkstand and a writing screen.

Bottle. China, nineteenth century. Height 3.3". American Museum of Natural History (Anthropology). Drummond Collection, 70.3.1907A

Both sides of this bottle are carved in shallow relief showing a pair of birds, one flying over the other. On one broad side are lotus flowers; on the other are stems with buds. The narrow sides have very similar portraits of a bald man with curled eyebrows and bared upper teeth, from which hangs a handle. The washer and spoon are of ivory.

Bottle. China, nineteenth century. Height 3.9". American Museum of Natural History (Anthropology). Drummond Collection, 70.3.1887A

Made of deep-red, transparent amber with a jade stopper, the bottle features two large peaches on a branch with leaves, the branch stretching along one side. The inner cavity has a very irregular wall and is remarkable in that the only opening for carving it is a hole at the top about one-eighth of an inch in diameter. The spatula attached to the stopper is of metal.

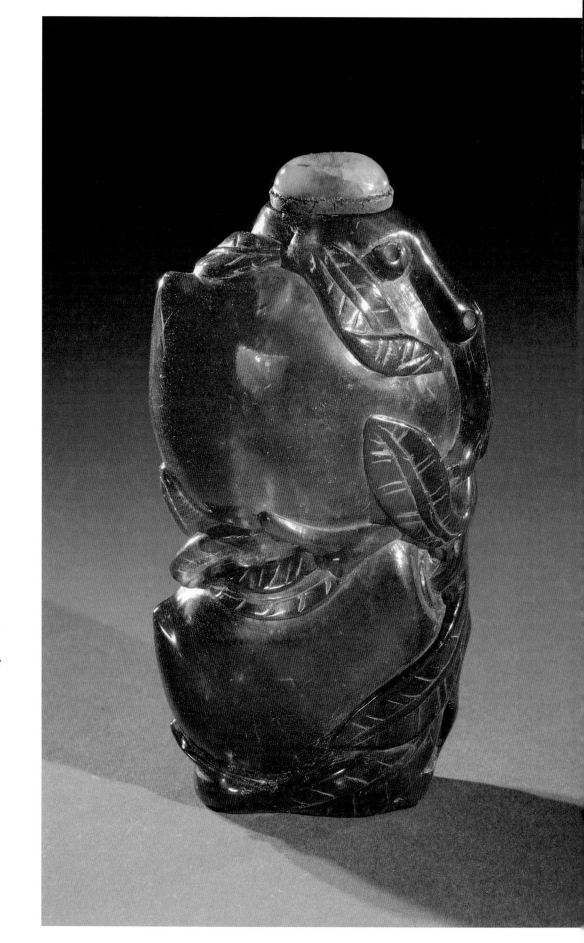

Bottle. China, nineteenth century. Height 3.5". American Museum of Natural History (Anthropology). Drummond Collection, 70.3.1922

Carved from an oval piece of opaque, yellowish, and semitranslucent reddish-brown amber, the bottle retains natural pits on the surface. The engraved lines are bounded by the color in the piece; the reverse side has script in imitation of an ancient style, with an undeciphered meaning.

Opposite and above: Two scenes. China, nineteenth century. Height 6.5″. American Museum of Natural History (Anthropology). Drummond Collection, 70.3.2672

Carved from a large piece of deep yellow-orange, transparent amber are two scenes in deep relief. On one side (opposite) a mother looks down on her child; they each hold a flower. Flowers and leaves are carved along the edges, and a large flower appears near the middle. The other side portrays a stag kneeling beside a gnarled tree in blossom. A peacock is perched in the tree, its tail curled over its back. A hole at the bottom indicates that the piece originally had a base.

Small bowl. China, eighteenth century. Length 4.8″. American Museum of Natural History (Anthropology). Drummond Collection, 70.3.2695

All around the bowl are entwined carved magnolia branches bearing leaves, flowers, and buds. The bowl itself is inscribed with a stylized flower. On the underside are buds, leaves, and a large flower as well as thick, gnarled branches with knots. The intricate work illustrates very effective use of the different colors occurring in a single amber piece.

Decorative "water pot." China, nineteenth century. Height 5.3″. American Museum of Natural History (Anthropology). Drummond Collection, 70.3.2671

The large, central, round object represents a water pot with the inscribed lid fixed to it. Surrounding the pot are branches, leaves, flowers, and peaches. Several darker pieces have been applied, such as the bat and the stem. The lighter bat was also applied. Bats are a welcome sign in China, emblematic of happiness and longevity. They are often rendered in red, as one is here, which is the color of joy. The other side is also carved, but without the bats.

Andrée, Karl. "Öst Preussens Bernstein und seine Bedeutung," *Östdeutscher Naturwart* S3 (1924): 120–34.

——. *Der Bernstein und seine Bedeutung in Natur und Geisteswissenschaften, Kunst und Kunstgewerbe Technik, Industrie und Handel.* Königsberg: Grfe und Unzer, 1937.

——. *Der Bernstein, das Bernsteinland und sein Leben.* Stuttgart: Kosmos, 1951.

Aschengreen Piacenti, Kristen. "Due Altari in Ambra al Museo degli Argenti," *Bollettino d'Arte,* 3–4 (July–December 1966): 163–66.

——. *Il Museo degli Argenti a Firenze.* Milan, 1968.

Bachofen-Echt, A. *Der Bernstein und seine Einschlüsse.* Wien: Springer-Verlag, 1949.

Bartel, M., and H. Hetzer. "Bernstein-Inklusen aus dem Miozän des Bitterfelder Raumes." *Zeitschrifte für angewandte Geologie* 28 (1982): 314–36.

Beck, Curt S., and Stephen Shennan. *Amber in Prehistoric Britain.* Oxbow Monograph 8. Oxford, 1991.

Beck, Curt S., M. Gerving, and E. Wilbur. "The Provenience of Archaeological Amber Artifacts." Parts 1 and 2. *Art and Archaeology Technical Abstracts* 4, nos. 2 and 3 (1966/1967): 215–302, 203–73.

Bellman, E. "Recovery and Treatment of Amber at Palmnicken (East Prussia)." *The Mining Journal* 102 (1913): 122.

Berendt, G. *Der im Bernstein befindlichen Organischen Reste Der Vorwelt.* Berlin, 1845.

Bradford, Ernle. *Four Centuries of European Jewelry.* Middlesex, England: Spring Books, 1953.

Buffum, Arnold. *The Tears of the Heliades, or Amber as a Gem.* London: Sampson Low, Marston and Co., 1897.

Chhibber, H. L. *The Mineral Resources of Burma.* London: Macmillan & Co., 1934.

Conwentz, Hugo. *Monographie der baltischen Bernsteinbäume.* Danzig, 1890.

Cook, John, M.D. *The Natural History of Lac, Amber, and Myrrh.* London, 1770.

DeNavarro, J. M. "Prehistoric routes between Northern Europe and Italy defined by the Amber Trade." *The Geographical Journal* 66 (1925): 481–507.

DeSalle, Rob, John Gatesy, Ward Wheeler, and David A. Grimaldi. "DNA Sequences from a Fossil Termite in Oligo-Miocene Amber and Their Phylogenetic Implications." *Science* 257 (1992): 1933–36.

Goeppert, H. R. *Über die Bernsteinflora.* Berlin: Monatsberichte Königisberg Academic der Wissenschaften, 1853.

——. *Die Flora des Bernsteins und ihre Beziehungen Zur Flora der Tertiärformation und der Gegenwart.* Vol. 1. Danzig, 1883.

Goeppert, H. R. and Berendt, G. C. *Die Bernstein und die in ihm befindlichen Pflan-zenreste der Vorwelt.* Vol. 1. Berlin, 1845.

Golenberg, Edward M. et al. "Chloroplast DNA Sequence from a Miocene *Magnolia* Species." *Nature* 344 (1990): 656–58.

Gough, L. J., and J. S. Mills. "The Composition of Succinite." *Nature* 239 (1972): 527–28.

Grabowska, J. *Amber in Polish History.* Edinburgh, 1978.

Grimaldi, David A., Curt W. Beck, and Jaap J. Boon. "Occurrence, Chemical Characteristics, and Paleontology of the Fossil Resins from New Jersey." *American Museum Novitates* 2948 (1989): 28 pp.

Grimaldi, David A., Elizabeth Bonwich, Michael Delannoy, and Stephen Doberstein. "Electron Microscopic Studies of Mummified Tissues in Amber Fossils." *American Museum Novitates* 3097 (1994): 31 pp.

Grimaldi, David A., Alexander Shedrinksy, Andrew Ross, and Norbert S. Baer. "Forgeries of Fossils in 'Amber': History, Identification, and Case Studies." *Curator* 37 (1994): 251–74.

Haddow, J. G. *Amber, All About It.* Cope's Smoke Room Booklets, no. 7. Liverpool, 1891.

Harding, A. F., and H. Hughes-Brock. "Amber in the Mycenaean World." *Annual of the British School at Athens* 69 (1974): 145–72.

Hartmann, P. J. *Succini Prussici physica et civilis historia.* Frankfurt, 1677.

Hennig, W. "*Fannia scalaris* Fabricius, eine rezente Art in Baltischen Bernstein? (Diptera: Muscidae)." *Stuttgarter Beitrage zur Naturkunde* (A) 150 (1966): 1–12.

Hong, Y.-Ch. et al. "Stratigraphy and Palaeontology of Fushun Coalfield, Liaoning Province." *Acta Geologica Sinica* 2 (1974): 113–49.

Hornaday, W. D. "Kauri Gum Deposits of New Zealand." *Mining Press* 110 (1915): 181–82.

Hueber, Francis M., and Jean L. Langenheim. "Dominican Amber Tree Had African Ancestors." *Geotimes* 31 (1986): 8–10.

Kornilowitch, Nicholas. "Has the Structure of Striated Muscle of Insects in Amber Been Preserved?" [In Russian.] *Prot. obschchestva estestro pri Itper. Yurev Univ.* 13 (1903): 198–206.

Lambert, J. B., C. W. Beck, and J. S. Frye. "Analysis of European Amber by Carbon-13 Nuclear Magnetic Resonance Spectroscopy." *Archaeometry* 30 (1988): 248–63.

Langenheim, J. L. "Amber, A Botanical Inquiry." *Science* 163 (1969): 1157–64.

Larsson, Sven G. *Baltic Amber: A Paleobiological Study.* Entomonograph 1. Klampenborg, Denmark: Scandinavian Science Press, 1978.

Laufer, Berthold. "Historical Jottings on Amber in Asia." *Memoirs of the American Anthropological Association* 1 (1907): 3.

Lee, Y.-T., and J. H. Langenheim. "Systematics of the Genus *Hymenaea* L. (Leguminosae, Caesalpinioideae, Detarieae). *University of California Publications in Botany* 69 (1975): 109 pp.

Ley, Willy. "The Story of Amber." *Natural History* (May 1938): 351–77.

MacAlpine, J. F. and J. E. H. Martin. "Canadian Amber—A Paleontological Treasure Chest." *The Canadian Entomologist* 101 (1969): 819–38.

Martinez, Ramon, and Dieter Schlee. "Die Dominikanischen Bernsteinminen der Nordkordillere, speziell auch aus der Sicht der Werkstätten." *Stuttgarter Beiträge zur Naturkunde* (C) 18 (1984): 78–84.

Michener, Charles D., and David A. Grimaldi. "A Trigona from Late Cretaceous Amber of New Jersey (Hymenoptera: Apidae: Meliponinae)." *American Museum Novitates* 2917 (1988): 10 pp.

Mustoe, G. E. "Eocene Amber from the Pacific Coast of North America." *Bulletin of the Geological Society of America* 96 (1985): 1530–36.

Objets d'Art in Amber from the Collection of the Catherine Palace Museum. Exhibition catalogue. Leningrad, 1990.

Pelka, Otto. *Die Meister der Bernsteinkunst.* Nuremberg, 1917.

——. *Bernstein.* Berlin: R. C. Schmidt, 1920.

——. "Zum Werk des Bernsteinmeisters Georg Schreiber in Königsberg," *Pantheon* 17 (1936): 27–29.

Penrose, R. A. F., Jr. "Kauri Gum Mining in New Zealand." *The Journal of Geology* 20 (1912): 1.

Pliny, the Elder, *The Natural History of Pliny.* Book 10, 37. Translated by D. E. Eichholz. Cambridge: Harvard University Press, 1962.

Poinar, George O., Jr. *Life in Amber.* Palo Alto: Stanford University Press, 1992.

Poinar, George O., Jr., and Roberta Hess. "Ultrastructure of 40-Million-Year-Old Insect Tissue." *Science* 215 (1982): 1241–42.

Prockat, Friedrich. "Amber Mining in Germany." *Engineering and Mining Journal* 129 (1932): 305–7.

Reineking von Bock, Gisela. *Bernstein, Das Gold der Ostsee.* Munich, 1981.

Rice, Patty C. *Amber, The Golden Gem of the Ages.* New York: The Kosciuszko Foundation, Inc., 1987.

Rohde, Alfred. *Bernstein, ein Deutscher Werkstoff. Seine Künstlerische Verarbeitung vom Mittelalter bis zum 18 Jahrhundert.* Berlin, 1937.

——. "Das Bernsteinzimmer Friedrichs I in Königsberger Schloss." *Pantheon* 29 (1942): 200–203.

Savkevich, S. S. "State of Investigation and Prospects for Amber in U.S.S.R." *International Geology Review* 17 (1975): 919–23.

Saunders, W. B., R. H. Mapes, F. M. Carpenter, and W. C. Elski. "Fossiliferous Amber from the Eocene (Claiborne) of the Gulf Coastal Plain." *Geological Society of America Bulletin* 85 (1974): 979–84.

Schlee, Dieter. "Bernstein aus dem Lebanon." *Kosmos* 68 (1972): 460–63.

——. *Bernstein-Raritäten.* Stuttgart: Staatliches Museum für Naturkunde, 1980.

——. "Bernstein-Neuigkeiten." *Stuttgarter Beiträge zur Naturkunde* (C) 18 (1984): 100 pp.

——. *Der Bernsteinwald.* Exhibition catalogue, Stuttgart: Staatliches Museum für Naturkunde, 1986.

——. "Das Bernstein-Kabinett." *Stuttgarter Beiträge zur Naturkunde* (C) 28 (1990): 100 pp.

Schlee, Dieter, and W. Glöckner. "Bernstein." *Stuttgarter Beiträge zur Naturkunde* (C) 8 (1978): 72 pp.

Schlüter, Thomas. "Zur Systematik und Palökologie harzkonservierter Arthropoda einer Taphozönose aus dem Cenomanium von NW-Frankreich." *Berliner Geowissenschaft Abhandlung* (A) 9 (1978): 150 pp.

Sendelio, Nathanael. *Historia Succinorum Corpora aliena involventium et Naturae Opere. . .* Elbing, 1742.

Shedrinsky, A. M., David A. Grimaldi, T. P. Wampler, and N. S. Baer. "Amber and Copal: Pyrolysis Gas Chromatographic (PyGC) Studies of Provenance." *Wiener Berichte Über Naturwissenschaft in der Kunst* 6–8 (1991): 37–63.

Shedrinsky, Alexander, M., David A. Grimaldi, J. J. Boon, and N. S. Baer. "Application of Pyrolysis Gas Chromatography and Pyrolysis Gas Chromatography / Mass Spectrometry to the Unmasking of Amber Forgeries." *Journal of Analytical and Applied Pyrolysis* 25 (1993): 77–95.

Spekke, Arnold. *The Ancient Amber Routes and the Geographical Discovery of the Eastern Baltic.* Stockholm: M. Goppers, 1957.

Strong, D. E. *Catalogue of the Carved Amber in the Department of Greek and Roman Antiquities.* London: The Trustees of the British Museum, 1966.

Trusted, Marjorie. *Catalogue of European Ambers in the Victoria and Albert Museum.* London: Trustees of the Victoria and Albert Museum, 1985.

Williamson, George C. *The Book of Amber.* London: Ernest Benn, 1932.

Wilson, Edward O., Frank M. Carpenter, and William L. Brown, Jr. "The First Mesozoic Ants." *Science* 157 (1967): 1038–40.

Zaddach, G. *Das Tertiärgebirge Samlands.* Schriften der Physikalische-Ökonomische Gesellschaft, Jg. 8. Königsberg, 1867.

Zherichin, V. V., and I. D. Sukacheva. "On Cretaceous Insect Bearing Ambers (Retinites) of Northern Siberia." [In Russian.] *Reports of the 24th Annual Lectures in Memory of N.A. Kholodkovsky,* pp. 3–48. Leningrad, 1973.

PHOTOGRAPH/ILLUSTRATION CREDITS

Academy of Natural Sciences, Philadelphia: 23, 39, 43

American Museum of Natural History:

Jacklyn Beckett: 2, 10, 18, 19, 21, 22, 26, 28, 38, 44, 48, 50, 54, 55, 62, 63, 70, 74, 75, 77, 78, 80, 81, 84, 86, 88, 92, 93, 94, 95, 96, 98, 103, 104, 107, 108, 110, 111, 131, 134

Denis Finnin: 7, 27, 33, 45, 51, 73, 100, 146, 150, 151, 153, 154 middle and bottom, 155, 159, 195, 196, 197, 198, 199, 200, 201, 203, 204, 205, 206, 207, 208

Edward Bridges: 67, 71, 83, 85, 86, 87, 89, 91, 102, 104, 106, 109, 115, 127, 135, 138, 139

David Grimaldi: 12, 13, 28, 31, 36, 37, 39, 65, 66, 69, 76, 77, 80, 82, 86, 93, 95, 97, 99, 100, 101, 112, 113, 114, 115, 118, 129, 135, 146

Elizabeth Bonwich/Peling Melville: 119, 120, 121, 123, 124, 125

Malcolm McKenna: 44

Library Services (Joel Sweimler): 17, 24, 49, 59, 60, 136, 146, 161, 162, 163

Matthew Harnick: 108

Booth Museum, Brighton: 147, 148

From Chhibber, H. L. *The Mineral Resources of Burma.* London: Macmillan & Co., 1934

Cornell University (William L. Crepet and Kevin Nixon): 29

Ekaterininsky Palace Museum, Saint Petersburg: 187

Stas V. Chabdutkin: 173, 176, 185, 188, 189, 192, 193

Kuji Amber Museum (Kazuhisa Sasaki): 32

Metropolitan Museum of Art: 152

Museum für Naturkunde, Stuttgart (Dieter Schlee): 38, 90

Museum of Comparative Zoology, Harvard (Frank Carpenter): 30, 56

Museum of Fine Arts, Boston: 6, 142, 165, 168, 169, 170, 171, 174, 175, 176, 177, 178 middle and bottom, 180, 181, 184

Nationalmuseet, Copenhagen: 145

The Natural History Museum, London: 41, 139, 183

Paleontological Institute, Moscow: J. Sukacheva: 34 top and middle A. Ponomarenko: 34 bottom, 35

Rolf Reinicke: 52, 58

Smithsonian Institution, National Museum of Natural History: 64

Victoria and Albert Museum, London: 172